INTERNATIONAL CENTRE FOR MECHANICAL SCIENCES

COURSES AND LECTURES - No. 309

TIME AND FREQUENCY REPRESENTATION OF SIGNALS AND SYSTEMS

EDITED BY

G. LONGO
UNIVERSITÀ DI TRIESTE

B. PICINBONO
LABORATOIRE DES SIGNAUX ET SYSTÈMES

SPRINGER-VERLAG WIEN GMBH

Le spese di stampa di questo volume sono in parte coperte da
contributi del Consiglio Nazionale delle Ricerche.

This volume contains 23 illustrations.

In order to make this volume available as economically and as
rapidly as possible the authors' typescripts have been
reproduced in their original forms. This method unfortunately
has its typographical limitations but it is hoped that they in no
way distract the reader.

ISBN 978-3-211-82143-5   ISBN 978-3-7091-2620-2 (eBook)
DOI 10.1007/978-3-7091-2620-2

PREFACE

This volume contains the text of four lectures given at the CISM from 21 to 25 September, 1987, under the general title "Time and frequency representation of signals and systems". We present our thanks to the administration of the CISM which made this meeting possible. It was an extremely pleasant week with many opportunities for meetings and discussions. We hope that this book will reflect the atmosphere of the lectures as well as their content.

The concept of time and frequency representation of signals is as old as the first notation of written music. If we consider the musical line of a single voice instrument such as a flute or oboe, we see a succession of notes which must be played in time. At each time instant the exact note being played can be detected. As each note is associated with a particular fundamental frequency, or eventually with some harmonic frequencies, we can say that the score is a time and frequency representation of a "signal", which is the whole piece of music.

Those interested in both music and physics can attempt to present a mathematical model of the phenomenon. For this purpose we have to find the appropriate theory, which is obviously the Fourier representation of signals and systems. Using this theory, we call $s(t)$ the signal representing the whole musical piece and, as the frequency aspect, at the basis of the theory of harmony, is fundamental, we calculate the Fourier transform $S(v)$ of this signal. But as this Fourier transform is completely independent of time, there is no way of finding which note is to be played at a particular time instant, although the ear makes this recognition very easily. In other words, there is a mathematical contradiction in talking about an instantaneous frequency, even though musical notation seems to use this concept.

The physicist cannot consider this situation as satisfactory, and as a result he must introduce some tools allowing the simultaneous representation of time and frequency. This is the basic theme of the lectures presented in this book.

In the first lecture the concept of the analytical signal is presented and discussed. It is the most satisfactory way of introducing the ideas of instantaneous amplitude and phase without ambiguity. The second lecture is devoted to time and frequency respresentations using bilinear structures. This is particularly appropriate for discussing energy distribution. The contradiction mentioned above is discussed in relation to the uncertainty principle introduced in quantum mechanics. The same kind of problem is discussed in the third lecture, but here in the case of stochastic signals. Finally, parametric methods are introduced in the last lecture. These methods are particularly appropriate in clarifying the relation between representations of signals and systems.

In spite of the various approaches outlined above, the book presents a unity in subject, and we hope that it will he helpful to scientists interested in the domain but perhaps confused by the great choice of literature in this field.

Bernard PICINBONO
Laboratoire des Signaux et Systèmes

CONTENTS

Page

CONTENTS

THE ANALYTICAL SIGNAL AND RELATED PROBLEM

B. Picinbono

Laboratoire des Signaux et Systèmes, Gif-sur-Yvette, France

ABSTRACT

In many communication systems it is necessary to use the concepts of instantaneous amplitude and phase of signals. These concepts are not naturally defined, and the classical Fourier theory shows a contradiction between the ideas of frequency and that of time. The classical way to overcome this difficulty is to use extensively the concept of the analytical signal. It is then possible to associate with any signal a pair of functions $[a(t), \phi(t)]$, which are instantaneous amplitude and phase. However, the analytical property introduces precise limitations, analyzed in this paper. The results of this analysis are particularly appropriate to phase signals, which are examined in detail. Some practical consequences are also discussed.

1. INTRODUCTION, HISTORICAL BACKGROUND

In many areas of signal processing it is important to write a real signal $x(t)$ in the form

$$x(t) = a(t) \cos[\omega_0 t + \Psi(t)] \tag{1.1}$$

which introduces the concepts of *instantaneous amplitude* $a(t)$, of *instantaneous phase* $\omega_0 t + \Psi(t)$, of *instantaneous frequency* $\omega_0 + d\Psi/dt$ and of mean angular frequency ω_0. These concepts are absolutely necessary in discussions of problems of amplitude modulation (AM) or frequency modulation (FM), which are used in everyday radiocommunication systems. But it is also well-known that the classical Fourier analysis describes a signal either in the time domain or in the frequency domain, which makes nonsense of the concept of instantaneous frequency. In the face of this contradiction between the purely theoretical approach and the demands of application, we should investigate the domain more carefully in order to arrive at a more satisfactory situation.

Writing (1.1) in the form $a(t) \cos[\phi(t)]$, we see that the problem is to associate to any real signal $x(t)$ a pair of functions $[a(t), \phi(t)]$ such that $x(t) = a(t) \cos[\phi(t)]$. It is obvious that a given pair defines $x(t)$ completely while a given signal $x(t)$ can be described by many different pairs. Thus, without further conditions, the definition of $a(t)$ and $\phi(t)$ from $x(t)$ is not a well-defined problem.

In order to remove this ambiguity, a classical method consists of the use of the *analytical signal*, also called the *complex envelope* of the signal. The idea is to associate with any real signal $x(t)$ a well-defined complex signal $z(t)$ such that its modulus and phase are $a(t)$ and $\phi(t)$ respectively. Conversely, we will call below a *canonical pair* a pair of functions $[a(t), \phi(t)]$ such that $z(t) = a(t) \exp[j\phi(t)]$ is analytical, as defined later. The use of the analytical signal is not recent [1 - 7] and has been justified by many physical arguments [7], including comparison with methods in quantum mechanics. It is not our purpose, however, to discuss this matter here.

When we discuss the properties of canonical pairs of functions, the first problem which appears is that of characterizing the pair. It is in fact well known [1] that the real and imaginary parts of an analytical signal are Hilbert transforms of eachother. Unfortunately there is no direct counterpart to this property concerning $a(t)$ and $\phi(t)$ except for minimum phase signals, which form a very limited class of signals. It is, of course, always possible to use the fact that $a(t) \cos[\phi(t)]$ and $a(t) \sin[\phi(t)]$ are Hilbert transforms, but this is a condition which is in general very difficult to verify. Furthermore, an analytical signal $z(t)$ is characterized completely by the fact that its Fourier transform (FT) $Z(v)$ vanishes for $v < 0$, which is a simple *spectral* property. It would thus be also very convenient to characterize a canonical pair only by spectral methods, and many attempts in this direction have been made [8 - 18]. One of the most usual is to use the Bedrosian theorem [12] concerning the Hilbert transform of the product of functions. More precisely, this states that if $a(t)$ is a real signal with an FT null outside the frequency band $[-B, +B]$ and if $\cos[\phi(t)]$ has the complementary property (FT null inside $[-B, +B]$), then

$$H\{a(t)\cos[\phi(t)]\} = a(t)\,H\{\cos[\phi(t)]\}. \tag{1.2}$$

But this does not prove that the Hilbert transform of $\cos[\phi(t)]$ is $\sin[\phi(t)]$, which is necessary to ensure that $a(t)\exp[j\phi(t)]$ is analytical. This problem has been investigated by many means [8 - 18], without significant success, and one of our objectives is to understand clearly the reasons for this. To this end we will study carefully the class of unimodular analytical signals which is the cornerstone of the whole discussion. First, let us recall some basic properties of analytical signals necessary to understand clearly the following argument.

2. CANONICAL PAIR - THE LIMITS OF THE FREQUENCY APPROACH

2.1. Definitions
A pair of functions $[a(t), \phi(t)]$ is canonical if and only if the signal $a(t)\exp[j\phi(t)]$ is analytical, which means that its FT vanishes for negative frequencies. This is equivalent to saying that the Hilbert transform of $a(t)\cos[\phi(t)]$ is $a(t)\sin[\phi(t)]$. Finally, it is worth noting that a complex signal $z(t)$ is analytical if for complex values of t the function $z(t)$ is analytic in the half upper complex plane, which means that there is no singularity in this half plane.

2.2. Application to amplitude modulation
Consider a purely monochromatic signal $x(t) = a\,\cos(\omega_0 t)$. The corresponding canonical pair is of course $[a, \omega_0 t]$ and the instantaneous amplitude is the constant a. Consider now the signal $m(t)\cos(\omega_0 t)$, sometimes referred to as the amplitude modulated signal. From this expression we could be tempted to conclude that its instantaneous amplitude is $m(t)$, or that the pair $[m(t), \omega_0 t]$ is canonical. This is, however, in general not true and to ensure this property, $m(t)$ must satisfy some specific condition. For amplitude modulation these conditions are very easy to find. In fact, $m(t)$ must be such that $m(t)\exp(j\omega_0 t)$ is analytical. As multiplication by $\exp(j\omega_0 t)$ generates a frequency translation in the Fourier domain, we conclude that $m(t)$ is the instantaneous amplitude of $m(t)\cos(\omega_0 t)$ if and only if $m(t)$ is band-limited in $[-B, +B]$ with $B = \omega_0/2\pi$. As a consequence it is clear that if this condition is not fulfilled, the pair $[m(t), \omega_0 t]$ is not a canonical pair, and the instantaneous amplitude of $m(t)\cos(\omega_0 t)$ is not $m(t)$, which may be surprising. Finally, note that the canonical pair has been characterized only by spectral considerations, but this is no longer the case for phase modulation.

2.2. Application to phase modulation

By translating the same ideas, it would be interesting if we could state that the instantaneous phase of $a \cos[\phi(t)]$ is $\phi(t)$. But the problem is much more complicated than that of amplitude modulation. In fact, this means that $H\{\cos[\phi(t)]\} = \sin[\phi(t)]$, where H is the Hilbert transform, a condition which requires very specific properties of $\phi(t)$, as analyzed below. This is of course equivalent to saying that $\exp[j\phi(t)]$ is analytic, and signals of this kind are called *pure phase signals*. At this point let us return to the use of the Bedrosian theorem to characterize a canonical pair. From frequency considerations only, we arrive at (1.2), but the canonical property requires a condition which is precisely that defining a pure phase signal. In other words, if $a(t)$ and $\cos[\phi(t)]$ satisfy the spectral conditions of the Bedrosian theorem, and if $\phi(t)$ is the phase of a pure phase signal, then the pair $[a(t), \phi(t)]$ is canonical. This leads us to a more careful study of the properties of pure phase signals.

3. PURE PHASE SIGNALS

The problem is to investigate the structure of $\phi(t)$ such that $[1, \phi(t)]$ is canonical or that $\exp[j\phi(t)]$ is analytical. This problem has been thoroughly discussed in the mathematical literature and we will present only the fundamental result without proof [19, 20]. The most general unimodular analytic signal can be written in the form

$$z(t) = B(t) \exp[j(\omega_0 t + \theta)], \tag{3.1}$$

where ω_0 and θ are arbitrary and $B(t)$ is a *Blaschke function* defined by

$$B(t) = \prod_{k=1}^{N} \frac{t - z_k}{t - z_k^*}, \quad \mathrm{Re}[z_k] > 0. \tag{3.2}$$

In this expression N is finite, but extensions for $N \to \infty$ are possible. It is obvious that $|z(t)| = 1$, which means that $z(t)$ is unimodular. The instantaneous phase of $z(t)$ is

$$\phi(t) = \theta + \omega_0 t + \Phi_B(t) \tag{3.3}$$

where $\Phi_B(t)$ is a Blaschke phase or the phase of $B(t)$. The most general canonical pair with constant amplitude is thus $[a, \theta + \omega_0 t + \Phi_B(t)]$. There is of course a large degree of freedom because of the number of parameters appearing in (3.2). However, this freedom is not complete, and any signal written as $a \cos[\phi(t)]$ is not necessarily a phase signal with a constant amplitude. To discuss this point we will present some of the major consequences of (3.3), which are necessary conditions that must be satisfied by a pure phase signal.

1. *A pure phase signal has one and only one spectral line.* This is a consequence of the structure of $B(t)$ which is a product of functions $B_k(t)$ and can be written as

$$B_k(t) = \frac{t - z_k}{t - z_k^*} = 1 - \frac{z_k - z_k^*}{t - z_k^*} . \tag{3.4}$$

It then follows that

$$B(t) = 1 + c(t) \tag{3.5}$$

where $c(t)$ is a signal of finite energy and with a continuous spectrum. Using (3.1) we find that $z(t)$ has a spectral line at the angular frequency ω_0, and that it is the only spectral line possible in its Fourier representation.

2. *The continuous part of the spectrum of $z(t)$ is to the right of its spectral line.* This is a direct consequence of the structure of $B(t)$ appearing in (3.2). In fact, all the poles of $B(t)$ are in the lower half complex plane and as a result the Fourier transform $\tilde{B}(v)$ of $B(t)$ is null for $v < 0$. Then $B(t)$ is an analytical signal with only a spectral line at the frequency $v = 0$ and after multiplication by $\exp(j\omega_0 t)$ we obtain the property 2.

3. *The analytic signal of a pure phase signal cannot be band-limited if it is not strictly monochromatic.* This last case appears if $B(t) = 1$ which makes $z(t)$ defined by (3.1) band limited. Let us thus assume that $B(t) \neq 1$. Using the form given in (3.5), we note that $c(t)$ is a rational function in t. It is well known that the FT of a rational function cannot be band limited. Another proof, without using this argument of rational function, can also be given [21].

4. *Structure of the phase of a pure phase signal.* The instantaneous phase of a pure phase signal is given by (3.3), and we will investigate further the Blaschke phase $\Phi_B(t)$. Using (3.2) we can write this phase in the form

$$\Phi_B(t) = \sum_{k=1}^{N} \phi_k(t) \quad \mathrm{mod}(2\pi) \tag{3.6}$$

where $\phi_k(t)$ is the phase corresponding to the term of the product (3.2) associated with the zero z_k. Writing this complex number as

$$z_k = a_k + j\, b_k, \quad b_k > 0, \tag{3.7}$$

we find

$$\phi_k(t) = 2\, tg^{-1}\left[a_k/(a_k - t)\right] . \tag{3.8}$$

As the phase is always defined modulo 2π, it is interesting to calculate its derivative for which this ambiguity disappears. This corresponds to the instantaneous angular frequency, and can be written as

$$\omega(t) = \omega_0 + 2 \sum_{k=1}^{N} \frac{b_k}{(a_k - t)^2 + b_k^2} \qquad (3.9)$$

The condition on b_k appearing in (3.7) ensures that $\omega(t) > \omega_0$, which is another presentation of the fact that the spectrum of $z(t)$ is located to the right of the spectral line defined by ω_0. Using the fact that N, a_k and b_k are arbitrary, with only the condition $b_k > 0$, the instantaneous frequency $\omega(t)$ can be described with a broad degree of freedom. The limitation on its structure, consequence of (3.9), have not yet been investigated in detail.

4. SOME CONSEQUENCES FOR SIGNAL REPRESENTATION

Let us now present the most direct consequences of the previous results on signal representation by instantaneous amplitude and phase.

1. The signal $a \cos[\phi(t)]$ has a constant instantaneous amplitude if and only if the function $\phi(t)$ has the form (3.3). In fact, if $\phi(t)$ is given by (3.3), the signal $\exp[j\phi(t)]$ is analytical, as is also $a \exp[j\phi(t)]$. As $a \cos[\phi(t)]$ is its real part, the instantaneous amplitude is a. Conversely, if that is true, the Hilbert transform of $\cos[\phi(t)]$ must be $\sin[\phi(t)]$, and $\exp[j\phi(t)]$ is then analytical, which gives (3.3).

2. It is in general impossible to characterize a canonical pair *by band limitation properties only*. To illustrate this point, let us discuss some examples. Consider the signal $x(t) = a(t) \cos[\phi(t)]$ and suppose that the FT of $a(t)$ is limited to $[-B, +B]$ while the FT of $\cos[\phi(t)]$ has the complementary property. We then deduce from the Bedrosian theorem that (1.2) is valid. But, in order to obtain a canonical pair $[a(t), \phi(t)]$, we must ensure that $a(t) \exp[j\phi(t)]$ is analytical, which is equivalent to the following:

$$H\{a(t) \cos[\phi(t)]\} = a(t) \sin[\phi(t)]. \qquad (4.1)$$

Using (1.2), this gives

$$H\{\cos[\phi(t)]\} = \sin[\phi(t)], \qquad (4.2)$$

which means that $[1, \phi(t)]$ is canonical. However, this again implies that $\phi(t)$ is given by (3.3). In other words, we need additional properties to deduce from the Bedrosian theorem that $[a(t), \phi(t)]$ is canonical.

Conversely, the conditions of the Bedrosian theorem are not necessary to obtain a canonical pair $[a(t), \phi(t)]$. To verify this point, consider the signal

$$z(t) = [\alpha - 2\pi j t]^{-2}. \tag{4.3}$$

It is obviously an analytical signal because of the localization of its poles. The corresponding instantaneous amplitude is

$$a(t) = [\alpha^2 + 4\pi^2 t^2]^{-1} \tag{4.4}$$

and the FT of this function is

$$A(\nu) = (2\alpha)^{-1} \exp[-\alpha|\nu|] \tag{4.5}$$

We then see that the instantaneous amplitude does not present any band limitation property.

3. Let us conclude with some practical considerations. We have seen in section 2.2 that the signal $m(t) \cos(\omega_0 t)$ has the instantaneous amplitude $m(t)$ if and only if $m(t)$ is strictly band-limited in $[-B, +B]$ with $B = \omega_0/2\pi$. In practice strictly band-limited signals are very rare, but it is clear that for low frequency signals the condition becomes asymptotically true when $\omega_0 \to \infty$. As for many practical signals $\omega_0 \gg 2\pi B$, we can say that $m(t)$ is in practice the instantaneous amplitude of $m(t) \cos(\omega_0 t)$. We can then state that the pair $[m(t), \omega_0 t]$ is asymptotically canonical. The questions discussed in this paper become more critical when the condition $\omega_0 \gg 2\pi B$ is not fulfilled, a situation which appears in some physical problems for which a more careful discussion is necessary.

The same situation arises for phase modulation. If (3.3) is not satisfied, the signal $\cos[\phi(t)]$ does not have a constant instantaneous amplitude. But this amplitude becomes asymptotically constant when $\omega_0 \to \infty$, a situation which can be considered realistic for many signals used in communication problems. This is, however, not always the case, and it is at those times that the present discussion becomes important.

REFERENCES

[1] PICINBONO, B., Principles of signals and systems, *Artech House*, London, 1988, p.47.

[2] GABOR, D., Theory of communications, *J. IEE*, Nov. 1946, **93**, pp.429-457.

[3] VILLE, J., Théorie et applications de la notion de signal analytique, *Cables et Trans-missions*, Jan. 1948, **3**, pp.61-74.

[4]   OSWALD, J., The theory of analytic band-limited signals applied to carrier systems, *IRE Trans. Comm. Theor.*, Dec. 1956, **3**, pp.244-251.

[5]   DUGUNDJI, J., Envelopes and pre-envelopes of real wave forms, *IRE Trans. Inf. Theor.*, March 1958, **4**, pp.53-57.

[6]   BEDROSIAN, E., The analytic signal representation of modulated waveforms, *Proc. IRE*, Oct. 1962, **50**, pp.2071-2076.

[7]   VAKMAN, D., On the definition of concepts of amplitude, phase and instantaneous frequency of a signal, *Sovietskoye Radio*, 1972, pp.754-759.

[8]   VOELCKER, H., Toward a unified theory of modulation. Part I, Phase envelope relationships, *Proc. IEEE*, March 1966, **54**, pp.340-353.

[9]   VOELCKER, H., Toward a unified theory of modulation. Part II, Zero manipulation, *Proc. IEEE*, May 1966, **54**, pp.737-755.

[10] LERNER, R.M., A matched filter detection system for complicated Doppler shifted signals, *IRE Trans. IT*, June 1960, **6**, pp.373-385.

[11] RUBIN, W.L., DI FRANCO, J.V., Analytic representation of wide-band radio frequency signals, *J. Franklin Inst.*, Mar. 1963, **275**, pp.197-204.

[12] BEDROSIAN, E., A product theorem for Hilbert transforms, *Proc. IEEE*, May 1963, **51**, pp.868-869.

[13] CHIOLLAZ, M., ESCUDIÉ, B., HELLION, A., Une condition nécessaire et suffisante pour l'écriture du modèle exponentiel des signaux d'énergie finie, *Ann. Télécomm.*, 1978, **33**, pp.69-70.

[14] RIHACZEK, A.W., BEDROSIAN, E., Hilbert transforms and the complex representation of real signals, *Proc. IEEE*, Mar. 1966, **54**, pp.434-435.

[15] NUTTAL, A.H., BEDROSIAN, E., On the quadrature approximation of the Hilbert transform of modulated signals, *Proc. IEEE*, Oct. 1966, **54**, pp.1458-1459.

[16] STARK, H., An extension of the Hilbert form product theorem, *Proc. IEEE*, Sep. 1971, **59**, pp.1349-1360.

[17] BEDROSIAN, E., STARK, H., Comments on "An extension of the Hilbert form product theorem", *Proc. IEEE*, Feb. 1972, **60**, pp.228-229.

[18] BROWN Jr., J.L., Analytic signals and product theorems for Hilbert transforms, *IEEE Trans., CAS*, Nov. 1974, **21**, pp.790-792.

[19] NEVANLINNA, R., Analytic functions, *Springer-Verlag*, Berlin, 1970, Chap. 7.

[20] EDWARDS, S.F., PARRENT, G.B., The form of the general unimodular analytic signal, *Optica Acta*, 1959, **6**, pp.367-371.

[21] PICINBONO, B., MARTIN, W., Représentation des signaux par amplitude et phase instantanées, *Ann. Télécommunications*, May 1983, pp.179-190.

* Laboratoire du CNRS, de l'École Supérieure d'Électricité et de l'Université de Paris-Sud.

A TUTORIAL ON NON-PARAMETRIC BILINEAR
TIME-FREQUENCY SIGNAL REPRESENTATIONS

W. Mecklenbräuker
Technical University of Vienna, Vienna, Austria

This chapter originally appeared in: J. L. Lacoume, T. S. Durrani and R. Stora, eds., Les Houches, Session XLV, 1985, Traitement du signal / Signal processing, © Elsevier Science Publishers B. V., 1987. Reprinted with permission.

Introduction

Nonstationary signals have a time-dependent spectral content. This is in contrast to stationary signals whose energy spectrum characterizes their spectral content and that is independent of time. Therefore, nonstationary signals require joint time–frequency representations.

One of the most popular time–frequency signal representations is the spectrogram, which is obtained from the short-time Fourier transform of the signal. Starting with a discussion of some properties of the spectrogram, the general class of bilinear signal representations (BSR) is introduced. It is indicated that all BSR's can be expressed by several different normal forms, each of which is characterized by a kernel function.

Imposing specific constraints on these kernel functions, many desirable properties of these BSR's can be obtained. In this way, BSR's can be equipped with properties that are important from the signal-processing point of view. Special emphasis is given to energy distributions (ED), whose properties are discussed in more detail. It is indicated that of the many possible ED's of a signal, the Wigner distribution (WD) has many advantages and is optimal in a certain sense.

A rather complete reference list with respect to the theory and applications of BSR's, especially the Wigner distribution, is included.

1. Remarks on the Fourier transform

The Fourier spectrum of a signal $x(t)$,

$$X(f) = \int_t x(t) e^{-j2\pi ft} dt = F_f\{x(t)\}. \tag{1.1}$$

characterizes the signal $x(t)$ uniquely, because $x(t)$ can be reconstructed from $X(f)$ by

$$x(t) = \int_f X(f) e^{j2\pi ft} df = F_t^{-1}\{X(f)\}. \tag{1.2}$$

$X(f)$ is thus the complex amplitude of the complex exponential $e^{j2\pi ft}$, which is used in eq. (1.2) to represent $x(t)$ for all values of t. Furthermore, $|X(f)|^2$ can be interpreted as the spectral energy density of the signal $x(t)$ since, by Parseval's theorem, the signal energy can be expressed by

$$E_x = \int_t |x(t)|^2 \, dt = \int_f |X(f)|^2 \, df. \tag{1.3}$$

However, $|X(f)|^2$ does not give any indication on the instantaneous energy distribution, because the exponentials in (1.2) are present for all times t.

A popular method to get a better idea of the time–frequency behaviour of the signal energy distribution is to consider the short-time Fourier transform of $x(t)$, which is discussed in the next section.

2. The short-time Fourier transform (STFT)

The STFT considers only specific time segments of the signal $x(t)$, which are obtained by applying a window function $w(t)$ to $x(t)$:

$$x_w(t_0, t) = x(t) w(t - t_0). \tag{2.1}$$

In general it is assumed that the window is real, has a finite duration D_w, and is centered at the instant t_0. The Fourier transform of a time segment of $x(t)$

$$X_w(t_0, f) = F_f\{x_w(t_0, t)\} = \int_t x(t) w(t - t_0) e^{-j2\pi ft} \, dt \tag{2.2}$$

again represents this segment for all values of t:

$$x_w(t_0, t) = \int_f X_w(t_0, f) e^{j2\pi ft} \, df. \tag{2.3}$$

However, because of eq. (2.1), it is additionally known that x_w is centered at t_0 and has duration D_w. Therefore the STFT gives a time–frequency representation of $x(t)$ if it is evaluated for all values of t_0. For further use, it can also be expressed by the Fourier spectra of $x(t)$ and $w(t)$ by convolution in the frequency domain

$$X_w(t_0, f) = \int_\eta X(\eta) W(f - \eta) e^{-j2\pi(f-\eta)t_0} \, d\eta. \tag{2.4}$$

3. Properties of the spectrogram

The squared modulus for the STFT, the spectrogram of $x(t)$,

$$S_x(t, f) = |X_w(t, f)|^2, \tag{3.1}$$

displays the signal energy density in time and in frequency. With eqs. (2.2) and (2.4), it can be expressed as

$$S_x(t, f) = \int_{t_1} \int_{t_2} x(t_1) x^*(t_2) w(t_1 - t) w(t_2 - t)$$

$$\times e^{-j2\pi f(t_1 - t_2)} \, dt_1 \, dt_2 \tag{3.2}$$

or

$$S_x(t, f) = \int_{\eta_1} \int_{\eta_2} X(\eta_1) X^*(\eta_2) W(f - \eta_1) W^*(f - \eta_2)$$

$$\times e^{j2\pi t(\eta_1 - \eta_2)} \, d\eta_1 \, d\eta_2. \tag{3.3}$$

Considering the time average of S_x

$$\int_t S_x(t, f) \, dt = \int_\eta |X(\eta)|^2 |W(f - \eta)|^2 \, d\eta \tag{3.4}$$

we see that this gives an average of the spectral energy density $|X(f)|^2$. This average coincides the better with $|X(f)|^2$ the better $|W(f)|^2$ approximates an impulse in frequency.

Likewise the frequency average of S_x

$$\int_f S_x(t, f) \, df = \int_t |x(\tau)|^2 w^2(\tau - t) \, d\tau \tag{3.5}$$

gives an average of the power of $x(t)$. The better $w^2(t)$ approximates an impulse function in time, the better this average gives the instantaneous behaviour of $x(t)$. While both averages (3.4) and (3.5) clearly show that a certain instantaneous energy distribution is approximated by the spectrogram the conflicting conditions on the window function $w(t)$ also show that an ideal behaviour can never be obtained by the spectrogram. Integrating S_x over the whole (t, f) plane yields

$$\int_t \int_f S_x(t, f) \, dt \, df = \int_\tau |x(\tau)|^2 \int_t w^2(\tau - t) \, dt \, d\tau$$

$$= E_x \cdot E_w \tag{3.6}$$

i.e. if the window has unit energy,

$$E_w = \int_t w^2(t) \, dt = \int_f |W(f)|^2 \, df = 1, \tag{3.7}$$

then this average of the spectrogram is equal to the signal energy E_x.

4. Quadratic forms of signals

Generalizing the concepts developed so far we observe that the spectro-gram of the signal $x(t)$ can also be considered as a special "quadratic" form of the signal $x(t)$:

$$T_x(t, f) = \int_{t_1}\int_{t_2} x(t_1)x^*(t_2)k_T(t, f; t_1, t_2)\, dt_1\, dt_2. \tag{4.1}$$

The kernel k_T characterizes this signal representation T_x completely, and for the spectrogram, it is given by

$$k_s(t, f; t_1, t_2) = w(t_1 - t)w(t_2 - t)e^{-j2\pi f(t_1 - t_2)}. \tag{4.2}$$

Expressing $x(t)$ by its Fourier transform in eq. (4.1) we obtain

$$T_x(t, f) = \int_{f_1}\int_{f_2} X(f_1)X^*(f_2)K_T(t, f; f_1, f_2)\, df_1\, df_2, \tag{4.3}$$

which is now a "quadratic" form of $X(f)$. The two kernels k_T and K_T are connected by a two-dimensional Fourier transform:

$$K_T(t, f; f_1, f_2) = \int_{t_1}\int_{t_2} k_T(t, f; t_1, t_2)e^{-j2\pi(f_1 t_1 - f_2 t_2)}\, dt_1\, dt_2. \tag{4.4}$$

Further examples of special signal representations T_x are given by:
—the signal power $p(t) = |x(t)|^2$ which is characterized by the kernel

$$k_p(t, f; t_1, t_2) = \delta(t - t_1)\,\delta(t - t_2) \tag{4.5}$$

—the spectral energy density $P(f) = |X(f)|^2$ with

$$K_P(t, f; f_1, f_2) = \delta(f - f_1)\,\delta(f - f_2). \tag{4.6}$$

—the total signal energy E_x with

$$k_E(t, f; t_1, t_2) = \delta(t_1 - t_2). \tag{4.7}$$

While all of these examples are more or less directly connected to the *energy* of the signal $x(t)$ there still exists another type of "quadratic" form, which is of the *correlation type*. For example the autocorrelation function of the signal $x(t)$ is given by

$$r_x(\tau) = \int_t x(t + \tau)x^*(t)\, dt. \tag{4.8}$$

This can be obtained from (4.1) by

$$k_r(\tau; t_1, t_2) = \delta(t_1 - t_2 - \tau). \tag{4.9}$$

Observe that the number and type of external variables of T_x depend on the specific signal characterization considered.

5. Cross-signal representations

So far the signal representation by the form T_x characterizes one single signal $x(t)$. However, consider the composite signal $z(t) = ax(t) + by(t)$, which is just a superposition of the signals x and y: then for expressing T_z in T_x and T_y also the cross representations T_{xy} and T_{yx} are needed:

$$T_z(t, f) = |a|^2 T_x + |b|^2 T_y + ab^* T_{xy} + a^* b T_{yx} \tag{5.1}$$

with

$$T_{xy}(t, f) = \int_{t_1} \int_{t_2} x(t_1) y^*(t_2) k_T(t, f; t_1, t_2) \, dt_1 \, dt_2 \tag{5.2}$$

and a corresponding expression for $T_{yx}(t, f)$. Without further properties of the kernel function $k_T(t, f; t_1, t_2)$ in general there does not exist any relation between T_{xy} and T_{yx}.

From eq. (5.2) it follows that T_{xy} is a linear functional in $x(t)$ and $y^*(t)$ respectively. From this property, all signal representations of the form (4.1) are called *bilinear*.

6. Normal forms of bilinear signal representations

So far the two normal forms (4.1) and (4.3) of bilinear signal representations T_x have been considered. Since the number and type of external variables are not important in this section, they will not explicitly be stated in T_x and its corresponding kernel functions. Thus the following notation is used in (4.1):

$$T_x(\cdot) = \int_{t_1} \int_{t_2} k_T(\cdot; t_1, t_2) x(t_1) x^*(t_2) \, dt_1 \, dt_2. \tag{4.1'}$$

Transformation of the integration variables in (4.1') by

$$t_1 = t + \tfrac{1}{2}\tau, \qquad t_2 = t - \tfrac{1}{2}\tau \tag{6.1}$$

yields

$$T_x(\cdot) = \int_t \int_\tau u_T(\cdot; t, \tau) x(t + \tfrac{1}{2}\tau) x^*(t - \tfrac{1}{2}\tau) \, dt \, d\tau \tag{6.2}$$

(Time domain normal form).

The new kernel u_T is connected to k_T by (6.1)

$$u_T(\cdot\,; t, \tau) = k_T(\cdot\,; t + \tfrac{1}{2}\tau, t - \tfrac{1}{2}\tau). \tag{6.3}$$

Similarly in (4.3) the integration variables are transformed by

$$f_1 = f + \tfrac{1}{2}\eta, \qquad f_2 = f - \tfrac{1}{2}\eta, \tag{6.4}$$

which yields

$$T_x(\cdot) = \int_f \int_\eta U_T(\cdot\,; f, \eta)\, X(f + \tfrac{1}{2}\eta)\, X^*(f - \tfrac{1}{2}\eta)\, df\, d\eta \tag{6.5}$$

(Frequency domain normal form).

U_T is obtained from K_T by

$$U_T(\cdot\,; f, \eta) = K_T(\cdot\,; f + \tfrac{1}{2}\eta, f - \tfrac{1}{2}\eta). \tag{6.6}$$

The two transformations (6.1) and (6.4) do not change the fundamental relationship (4.4), i.e. u_T and U_T are also a two-dimensional Fourier transform pair:

$$U_T(\cdot\,; f, \eta) = \int_t \int_\tau u_T(\cdot\,; t, \tau)\, e^{j2\pi(f\tau + \eta t)}\, dt\, d\tau. \tag{6.7}$$

Next two additional normal forms of T_x are obtained in the following way. Expressing the kernal u_T by its inverse Fourier transform with respect to τ

$$u_T(\cdot\,; t, \tau) = \int_f v_T(\cdot\,; t, f)\, e^{-j2\pi f\tau}\, df \tag{6.8}$$

and inserting this into eq. (6.2) we get the representation

$$T_x(\cdot) = \int_t \int_f \left[\int_\tau x(t + \tfrac{1}{2}\tau) x^*(t - \tfrac{1}{2}\tau)\, e^{-j2\pi f\tau}\, d\tau \right] v_T(\cdot\,; t, f)\, dt\, df. \tag{6.9}$$

The term in brackets is the Wigner Distribution (WD) of the signal $x(t)$:

$$W_x(t, f) = \int_\tau x(t + \tfrac{1}{2}\tau) x^*(t - \tfrac{1}{2}\tau)\, e^{-j2\pi f\tau}\, d\tau. \tag{6.10}$$

With eqs. (6.10) and (6.9) any bilinear signal representation can also be expressed by

$$T_x(\cdot) = \int_t \int_f v_T(\cdot\,; t, f)\, W_x(t, f)\, dt\, df \tag{6.11}$$

(Wigner distribution normal form).

Similarly u_T can be expressed by its Fourier transform with respect to t,

$$u_T(\cdot\,; t, \tau) = \int_\eta V_T(\cdot\,; \tau, \eta)\,e^{-j2\pi\eta t}\,d\eta \tag{6.12}$$

and with (6.2) this gives

$$T_x(\cdot) = \int_\tau \int_\eta \left[\int_t x(t + \tfrac{1}{2}\tau)x^*(t - \tfrac{1}{2}\tau)\,e^{-j2\pi\eta t}\,dt \right] V_T(\cdot\,; \tau, \eta)\,d\tau\,d\eta. \tag{6.13}$$

The term in brackets is the ambiguity function

$$A_x(\tau, \eta) = \int_t x(t + \tfrac{1}{2}\tau)x^*(t - \tfrac{1}{2}\tau)\,e^{-j2\pi\eta t}\,dt. \tag{6.14}$$

With this result the fourth normal form of a bilinear signal representation is obtained

$$T_x(\cdot) = \int_\tau \int_\eta V_T(\cdot\,; \tau, \eta)\,A_x(\tau, \eta)\,d\tau\,d\eta \tag{6.15}$$

(Ambiguity function normal form).

While all of these normal forms of a BSR have been derived from the time domain normal form (6.2) they are also obtainable from the frequency domain normal form (6.5).

Then the Wigner distribution and the ambiguity function are expressed by the spectrum of $x(t)$:

$$W_x(t, f) = \int_\eta X(f + \tfrac{1}{2}\eta)X^*(f - \tfrac{1}{2}\eta)\,e^{j2\pi\eta t}\,d\eta, \tag{6.16}$$

$$A_x(\tau, \eta) = \int_f X(f + \tfrac{1}{2}\eta)X^*(f - \tfrac{1}{2}\eta)\,e^{j2\pi f\tau}\,df. \tag{6.17}$$

The relations (6.7), (6.8) and (6.12) between the different kernel functions are depicted in the following diagram, where a Fourier transform pair is indicated by an arrow.

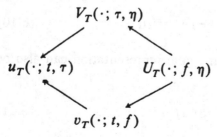

Similar relationships hold for the Wigner distribution, the ambiguity function, the time signal product

$$q_x(t, \tau) = x(t + \tfrac{1}{2}\tau)x^*(t - \tfrac{1}{2}\tau) \tag{6.18}$$

and the frequency spectrum product

$$Q_x(f, \eta) = X(f + \tfrac{1}{2}\eta)X^*(f - \tfrac{1}{2}\eta), \tag{6.19}$$

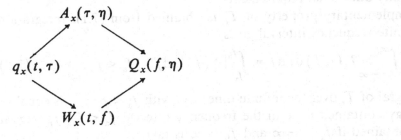

7. Energy distributions

A bilinear time frequency signal representation $T_x(t, f)$ will be considered as an energy distribution (ED) if its time average is equal to the spectral energy density,

$$\int_t T_x(t, f)\, dt = |X(f)|^2, \tag{7.1}$$

and its frequency average is equal to the signal power,

$$\int_f T_x(t, f)\, df = |x(t)|^2. \tag{7.2}$$

From the discussion in section 3 it should be clear that the spectrogram is not an energy distribution in the sense of eqs. (7.1) and (7.2). However, it approximates an energy distribution in a way that will be clarified later.

Although it is also possible to consider additionally signal representations $T_x(\tau, \eta)$ that are correlation representations and are Fourier transforms of the ED's (see Hlawatsch 1985), this aspect will not be pursued in this course.

From (7.2) it follows that the signal energy contained in the time interval $t_a < t < t_b$ is given by

$$\int_{t_a}^{t_b}\int_{f=-\infty}^{\infty} T_x(t, f)\, df\, dt = \int_{t_a}^{t_b}|x(t)|^2\, dt, \quad t_a < t_b. \tag{7.3}$$

Therefore this energy is distributed over the infinite frequency strip with

$t_a < t < t_b$. The total energy in $x(t)$ is given by the integral of T_x over the whole plane (t, f):

$$\int_t \int_f T_x(t, f)\, dt\, df = \int_t |x(t)|^2\, dt = E_x. \qquad (7.4)$$

Thus in contrast to the spectrogram, T_x gives the total signal energy without any additional requirements.

A complementary property of T_x is obtained from (7.1). Integration over a finite frequency interval gives

$$\int_{f_a}^{f_b} \int_{t=-\infty}^{\infty} T_x(t, f)\, dt\, df = \int_{f_a}^{f_b} |X(f)|^2\, df, \quad f_a < f_b. \qquad (7.5)$$

The integral of T_x over the infinite time strip with $f_a < f < f_b$ is equal to the energy contained in x in the frequency interval $f_a < f < f_b$. Again (7.4) is obtained if $f_a = -\infty$ and $f_b = \infty$ is taken.

The normal forms for T_x can now be used to derive the conditions on the corresponding kernel functions such that T_x is an ED:

Equation (7.1) is guaranteed only if the kernel v_T satisfies the condition

$$\int_t v_T(t, f; t', f')\, dt = \delta(f - f'). \qquad (7.6)$$

Similarly (7.2) requires that

$$\int_f v_T(t, f; t', f')\, df = \delta(t - t') \qquad (7.7)$$

is fulfilled.

From eq. (6.11) it is obvious that the WD is characterized by the kernel

$$v_W(t, f; t', f') = \delta(t - t')\, \delta(f - f') \qquad (7.8)$$

which fulfills both conditions (7.6) and (7.7). Therefore, the WD is an energy distribution.

The kernel v_A of the ambiguity function is given by

$$v_A(t, f; t', f') = e^{j2\pi(f't - t'f)} \qquad (7.9)$$

and does not comply with (7.6) and (7.7). The ambiguity function is not an energy distribution, rather it is a correlation representation (Hlawatsch 1985).

For a nice interpretation of $T_x(t, f)$ it is desirable to have ED's which are real-valued and non-negative.

While

T_x is real for all signals $x(t)$ (7.10)

if its kernel u_T is conjugate symmetric in the variable τ':

$$u_T(t; f; t', \tau') = u_T^*(t, f; t', -\tau'),$$ (7.11)

non-negativeness cannot be assured for bilinear energy distributions. This fact has been proved by Wigner (1979).

At first sight, this result might be disappointing because of the difficulty to interpret a negative value of an energy density. However, one should realize that Heisenberg's uncertainty relation holds between the duration of a signal and its occupied bandwidth in the frequency domain. This forbids to fix time and frequency simultaneously and arbitrarily sharply. Therefore, in any case, whether or not T_x is non-negative for all signals $x(t)$, a *pointwise* interpretation of an ED is always to be taken with care. A more detailed discussion of this sometimes difficult point will be given later.

8. Time- and frequency-shift properties

ED's should have two additional properties such that time shifts of the signal and shifts of the spectrum are properly represented:

If the signal $x(t)$ is shifted in time by t_0,

$$\mathscr{S}_{t_0} x(t) = x(t - t_0),$$ (8.1)

the corresponding ED T_x should also be shifted in time by t_0:

$$T_{\mathscr{S}_{t_0} x}(t, f) = T_x(t - t_0, f).$$ (8.2)

Similarly, if a signal $x(t)$ modulates a carrier signal $e^{j2\pi f_0 t}$,

$$\mathscr{M}_{f_0} x(t) = x(t) e^{j2\pi f_0 t},$$ (8.3)

its spectrum is shifted in frequency by f_0

$$\mathscr{S}_{f_0} X(f) = X(f - f_0)$$

and the corresponding ED T_x should then also be shifted in frequency by f_0:

$$T_{\mathscr{M}_{f_0} x}(t, f) = T_x(t, f - f_0).$$ (8.4)

Combining the shifts in time and in frequency yields the delayed and modulated signal

$$\mathscr{M}_{f_0} \mathscr{S}_{t_0} x(t) = x(t - t_0) e^{j2\pi f_0 t}.$$ (8.5)

Investigating the WD of $\mathcal{M}_{f_0} \mathcal{S}_{t_0} x$ we find that

$$W_{\mathcal{M}_{f_0} \mathcal{S}_{t_0} x}(t, f) = W_x(t - t_0, f - f_0), \tag{8.6}$$

i.e. the WD is time–frequency shift invariant.

Applying this result to the normal form (6.11), we get

$$T_{\mathcal{M}_{f_0} \mathcal{S}_{t_0} x}(t, f) = \int_{t'} \int_{f'} W_{\mathcal{M}_{f_0} \mathcal{S}_{t_0} x}(t', f') v_T(t, f; t', f') \, dt' \, df' \tag{8.7}$$

$$= \int_{t'} \int_{f'} W_x(t', f') v_T(t, f; t' + t_0, f' + f_0) \, dt' \, df' \tag{8.8}$$

with an obvious transformation of the integration variables.

If T_x has to be time–frequency shift invariant, (8.8) should be identical to

$$T_x(t - t_0, f - f_0) = \int_{t'} \int_{f'} W_x(t', f')$$
$$\times v_T(t - t_0, f - f_0; t', f') \, dt' \, df'. \tag{8.9}$$

Comparison of eqs. (8.8) and (8.9) yields the condition

$$v_T(t - t_0, f - f_0; t', f') = v_T(t, f; t' + t_0, f' + f_0) \tag{8.10}$$

which has to hold for all values of t_0 and f_0.

Taking $t' = 0$ and $f' = 0$ this gives

$$v_T(t - t_0, f - f_0; 0, 0) = v_T(t, f; t_0, f_0) \tag{8.11}$$

which means that v_T has to have the special form

$$v_T(t, f; t', f') = \psi_T(t - t', f - f'). \tag{8.12}$$

Thus the normal form of time–frequency shift invariant bilinear signal representations is given by

$$C_x(t, f) = \int_{t'} \int_{f'} W_x(t', f') \, \psi_c(t - t', f - f') \, dt' \, df' \tag{8.13}$$

C_x is a two-dimensional convolution of the WD and the kernel function ψ_c in the (t, f) plane. These shift-invariant signal representations are sometimes called after Cohen (1966).

C_x is an ED if ψ_c fulfills the conditions

$$\int_t \psi_c(t, f) \, dt = \delta(f) \tag{8.14}$$

and

$$\int_f \psi_c(t, f)\, df = \delta(t). \tag{8.15}$$

Besides the Wigner distribution the spectrogram S_x is also a member of the shift-invariant class, however, S_x is not an ED. By taking the corresponding Fourier transforms of ψ_c also the other kernels of the normal forms in section 6 can be determined for the shift invariant class C_x:

$$C_x(t, f) = \int_{t'}\int_{\tau'} \varphi_c(t - t', \tau')\, e^{-j2\pi f\tau'}\, q_x(t', \tau')\, dt'\, d\tau' \tag{8.16}$$

time domain normal form,

$$C_x(t, f) = \int_{\tau}\int_{\eta} \Psi_c(\tau, \eta)\, e^{j2\pi(t\eta - f\tau)}\, A_x(\tau, \eta)\, d\tau\, d\eta, \tag{8.17}$$

ambiguity function normal form,

$$C_x(t, f) = \int_{f'}\int_{\eta} \Phi_c(f - f', \eta)\, e^{j2\pi t\eta}\, Q_x(f', \eta)\, df'\, d\eta. \tag{8.18}$$

frequency domain normal form.

The four new kernels ψ_c, φ_c, Ψ_c, Φ_c are again related by the Fourier transform as shown in the following diagram:

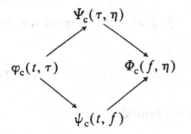

9. Examples of ED's

Discuss the properties of the following time–frequency signal representations and energy distributions:

(1) $R_x(t, f) = x^*(t)\, X(f)\, e^{j2\pi ft}$ Rihaczek (1968);

(2) $\mathrm{Re}\{R_x(t, f)\}$;

(3) $S_x^-(t, f) = |X^-(t, f)|^2$
   past time spectrogram,
   with $X^-(t, f) = \int_{-\infty}^t x(\tau)\, e^{-j2\pi f\tau}\, d\tau$;

(4) $L_x^-(t, f) = \dfrac{\partial}{\partial t} S_x^-(t, f)$ Page (1952);

(5) $S_x^+(t, f) = |X^+(t, f)|^2$

    future time spectrogram,

    with $X^+(t, f) = \displaystyle\int_{\tau=t}^{\infty} x(\tau)\, e^{-j2\pi f \tau} d\tau;$

(6) $L_x^+(t, f) = -\dfrac{\partial}{\partial t} S_x^+(t, f)$ Levin (1964).

(5) Derive the relationship between the representation (2), L_x^- and L_x^+.

(7) $\dfrac{1}{E_x} |x(t)|^2 |X(f)|^2.$

10. Uncertainty relation for ED's

In this section, it will be shown that the global characterization of ED's by the averages (7.1) and (7.2) already implies a certain local behavior of all ED's.

The key to this result is Heisenberg's uncertainty relation. Expressing the duration of a signal $x(t)$ with respect to an arbitrary time instant t_0 by

$$D_x^2 = \frac{2\pi}{E_x} \int_t (t - t_0)^2 |x(t)|^2 \, dt \tag{10.1}$$

and the frequency bandwidth occupied by $x(t)$ with respect to an arbitrary frequency f_0 by

$$B_x^2 = \frac{2\pi}{E_x} \int_f (f - f_0)^2 |X(f)|^2 \, df \tag{10.2}$$

then the duration-bandwidth product is lower bounded by

$$D_x B_x \geqslant \tfrac{1}{2}. \tag{10.3}$$

Usually t_0 in (10.1) and f_0 in (10.2) can be used to minimize D_x and B_x individually. However, in this context this will not be done.

With (7.1) and (7.2) signal duration and occupied bandwidth can also be expressed by the second moments of the ED T_x with respect to time and frequency:

$$D_x^2 = \frac{2\pi}{E_x} \int_t \int_f (t - t_0)^2 \, T_x(t, f) \, dt \, df, \tag{10.4}$$

$$B_x^2 = \frac{2\pi}{E_x} \int_t \int_f (f - f_0)^2 \, T_x(t, f) \, dt \, df. \tag{10.5}$$

Furthermore, the uncertainty relation (10.3) can be brought into the form

$$\frac{1}{T^2}D_x^2 + T^2 B_x^2 \geq 2D_x B_x \geq 1, \tag{10.6}$$

where the additional parameter $T > 0$ can be chosen freely. With eqs. (10.4) and (10.5) this yields the uncertainty relation for time–frequency ED's:

$$2\pi \int_t \int_f \left[\frac{(t - t_0)^2}{T^2} + T^2(f - f_0)^2 \right] T_x(t, f)\, dt\, df$$

$$\geq \int_t \int_f T_x(t, f)\, dt\, df = E_x. \tag{10.7}$$

This relation means that any ED T_x cannot be totally concentrated in an ellipse with semiaxes $(T/\sqrt{2\pi}, 1/T\sqrt{2\pi})$ in the (t, f) plane centered at any point (t_0, f_0).

This 'Heisenberg ellipse' has an area $\frac{1}{2}$. Expressed loosely this result shows that T_x must have a minimum spread corresponding to an area $\frac{1}{2}$ in the (t, f) plane. It should be stressed that this result holds for all ED's, i.e. as long as (7.1) and (7.2) are satisfied.

11. Finite support properties of ED's

It is easy to show that the non-negative ED's vanish for those values of t and f where the signal power or the spectral energy density are zero. This means that for non-negative ED's the signal energy is concentrated in those time or frequency strips which correspond to nonzero signal power or spectral energy density. For example, the spectrogram does not possess this property, which makes it difficult to interpret this signal representation correctly. As already remarked, bilinear time–frequency ED's can have negative values and therefore this type of concentration of T_x must be additionally imposed. More specifically, it is required that for signals with a finite extent (support) in time or frequency, the corresponding ED's should have the same finite support in the corresponding variables.

Hence, if $x(t)$ is a time-limited signal, i.e.

$$x(t) = 0 \quad \text{for} \quad t \notin T = [t_a, t_b],$$

then

$$C_x(t, f) = 0 \quad \text{for} \quad t \notin T. \tag{11.1}$$

Similarly, if $x(t)$ is a band-limited signal, i.e.

$$X(f) = 0 \quad \text{for} \quad f \notin F = [f_a, f_b]$$

then

$$C_x(t, f) = 0 \quad \text{for} \quad f \notin F. \tag{11.2}$$

It can be shown that ED's with these finite-support properties must have kernel functions that are constrained by

$$u_T(t, f; t', \tau') = 0 \quad \text{for} \quad |t - t'| > \tfrac{1}{2}|\tau'| \tag{11.3}$$

for the finite-support property in time, while

$$U_T(t, f; f', \eta') = 0 \quad \text{for} \quad |f - f'| > \tfrac{1}{2}|\eta'| \tag{11.4}$$

must hold for the finite-support property in frequency. Observe that the kernels of the Wigner distribution,

$$u_w(t, f; t', \tau') = \delta(t - t') e^{-j2\pi\tau'f}, \tag{11.5}$$

$$U_W(t, f; f', \eta') = \delta(f - f') e^{j2\pi\eta't}, \tag{11.6}$$

fulfill the corresponding conditions (11.3) and (11.4), so that the WD has the finite-support properties in time and in frequency.

As a consequence of these finite-support properties, for signals x that vanish for all $t < 0$ (this is true for impulse responses of causal linear time-invariant systems) the ED vanishes for all $t < 0$ as well:

$$T_x(t, f) = 0 \quad t < 0. \tag{11.7}$$

Similarly, for analytic signals x, whose spectrum vanishes for $f < 0$, the corresponding ED vanishes for $f < 0$:

$$T_x(t, f) = 0 \quad f < 0. \tag{11.8}$$

It will turn out that eq. (11.7) is important if linear time-invariant systems are characterized by ED's and that eq. (11.8) is helpful because it partially suppresses certain undesirable effects, namely the occurrence of interference terms in ED's.

12. Local moments of ED's and their relation to desirable signal characterizations

While the finite-support properties of the foregoing section restrict the form of an ED only very globally there still exist very different ways in which the global averages (7.1) and (7.2) can be satisfied by a bilinear signal representation T_x.

However, for certain signals, the desirable local behaviour of an ED can be described more specifically. Take for example a signal of the form

$$x(t) = a(t)\, e^{j\varphi(t)} \quad \text{with } a(t) \geqslant 0. \tag{12.1}$$

Under mild conditions on the envelope $a(t)$ and the phase $\varphi(t)$, the instantaneous behaviour of $x(t)$ in a time interval centered at t_0 is completely described by

$$z(t) = a(t)\, e^{j[\varphi'(t_0)\, t + \varphi_0]}. \tag{12.2}$$

$z(t)$ is a complex exponential with instantaneous frequency

$$f(t_0) = \frac{1}{2\pi}\varphi'(t_0) \tag{12.3}$$

at $t = t_0$.

It would be very desirable if for a signal of the form (12.1) the ED would be concentrated around the point $(t_0, f(t_0))$ in the (t, f) plane. This property can be expressed as follows. The normalized first-order moment, which can be interpreted as the centre of gravity in frequency, should be equal to the instantaneous frequency:

$$\frac{\int_f f\, T_x(t, f)\, df}{\int_f T_x(t, f)\, df} = \frac{1}{2\pi}\varphi'(t) = \frac{1}{2\pi}\,\mathrm{Im}\left\{\frac{x'(t)}{x(t)}\right\}. \tag{12.4}$$

Since T_x is assumed to be an ED the denominator of eq. (12.4) is equal to the power of $x(t)$: $|x(t)|^2 = a^2(t)$. For ED's the kernel U_T should then additionally satisfy the constraint

$$\int_f f\, U_T(t, f; f', \eta)\, df = f'\, e^{j2\pi\eta t} \tag{12.5}$$

or if we restrict T_x to belong to the Cohen class the constraint on $\psi_c(t, f)$ is given by

$$\int_f f\, \psi_c(t, f)\, df = 0. \tag{12.6}$$

Since for the Wigner distribution the kernel ψ_w is given by

$$\psi_w(t, f) = \delta(t)\, \delta(f) \tag{12.7}$$

it is seen that the WD has indeed the desired property (12.4).

Examples:

$$x(t) = A\,e^{j2\pi f_0 t}: \qquad W_x(t, f) = |A|^2\,\delta(f - f_0), \tag{12.8}$$

$$x(t) = A\,e^{j2\pi\alpha t^2/2}: \qquad W_x(t, f) = |A|^2\,\delta(f - \alpha t). \tag{12.9}$$

Observe that, in both cases, the WD is only concentrated on the line $(t, f(t))$ in the (t, f) plane. Especially the last signal is very remarkable: its instantaneous frequency is a linear function of time. The WD displays the instantaneous behaviour of this chirp signal very clearly. This is in contrast to its Fourier transform

$$X(f) = A\sqrt{\frac{j}{\alpha}}\,e^{-j\frac{\pi f^2}{\alpha}}, \tag{12.10}$$

which has constant modulus for all f. $|X(f)|^2$ only indicates that generally speaking, 'all' frequencies are present in $x(t) = A\,e^{j2\pi\alpha t^2/2}$.

While the instantaneous-frequency property (12.4) works well for complex-valued signals as in (12.1), it does not give the desired results for real-valued signals $x(t)$, since the right-hand side of (12.4) is equal to zero for real-valued $x(t)$. This problem is solved by representing real-valued signals by their corresponding complex-valued analytic signals defined by

$$z(t) = x(t) + j\breve{x}(t). \tag{12.11}$$

The imaginary part of $z(t)$ is the Hilbert transform of $x(t)$

$$\breve{x}(t) = \frac{1}{\pi}\int_\tau \frac{x(\tau)}{t - \tau}\,d\tau \tag{12.12}$$

and its Fourier transform is given by

$$\breve{X}(f) = -j\,\mathrm{sgn}\,f\,X(f). \tag{12.13}$$

Therefore, the spectrum of $z(t)$ has the simple form

$$Z(f) = \begin{cases} 2X(f) & f > 0, \\ 0 & f < 0, \end{cases} \tag{12.14}$$

and contains the same spectral information as $x(t)$.

Therefore if for real-valued signals the ED should display the instantaneous frequency, the ED is considered for the corresponding analytic signals $z(t)$.

A property similar to (12.4) can also be required for the localization of the centre of gravity in time of an ED.

Let us write the spectrum of $x(t)$ as

$$X(f) = A(f) e^{j\Phi(f)} \quad \text{with} \quad A(f) \geqslant 0. \tag{12.15}$$

The derivative of the phase $\Phi(f)$ is important, if $X(f)$ is the transmission function of a linear time invariant system. This can be seen as follows: the system is excited by an input signal

$$z(t) = v(t) e^{j2\pi f_0 t} \tag{12.16}$$

whose spectrum

$$Z(f) = V(f - f_0) \tag{12.17}$$

occupies only a small frequency band centered around f_0. The output signal $y(t)$ of the system has a spectrum given by

$$Y(f) = X(f) \cdot Z(f). \tag{12.18}$$

Again, as in (12.2), we can express $X(f)$ as

$$X(f) = A(f_0) e^{j\Phi(f_0)} \cdot e^{j[f-f_0]\Phi'(f_0)} \tag{12.19}$$

if the amplitude $A(f)$ is essentially constant in the frequency band considered and the phase $\Phi(f)$ may be approximated by the first two terms of its Taylor series expansion. With (12.17) and (12.19), $Y(f)$ is given by

$$Y(f) = A(f_0) e^{j\Phi(f_0)} \cdot V(f - f_0) e^{j[f-f_0]\Phi'(f_0)} \tag{12.20}$$

and the output signal $y(t)$ results in the form

$$y(t) = v\big(t - \tau_g(f_0)\big) \cdot A(f_0) e^{j2\pi f_0(t - \tau_p(f_0))}. \tag{12.21}$$

Here

$$\tau_g(f_0) = -\frac{1}{2\pi} \Phi'(f_0) \tag{12.22}$$

is the group delay or the delay of the (low-frequency) envelope $v(t)$ caused by the system and

$$\tau_p(f_0) = -\frac{1}{2\pi} \frac{\Phi(f_0)}{f_0} \tag{12.23}$$

is the delay of the (high-frequency) carrier of the input. The group delay can therefore very easily be interpreted as the delay of the signal energy at frequency f_0 and should be recognized in the ED.

This discussion leads to the required property: the normalized first-order moment in time should be equal to the group delay,

$$\frac{\int_t t T_x(t, f) \, dt}{\int_t T_x(t, f) \, dt} = -\frac{1}{2\pi} \Phi'(f_0) = -\frac{1}{2\pi} \text{Im}\left\{\frac{X'(f)}{X(f)}\right\}. \qquad (12.24)$$

The constraint on the kernel u_T such that the group delay property of an ED results in

$$\int_t t \, u_T(t, f; t', \tau') \, dt = t' e^{-j2\pi\tau'f}. \qquad (12.25)$$

For ED's of the Cohen class, the corresponding requirement is

$$\int_t t \psi_c(t, f) \, dt = 0. \qquad (12.26)$$

With (12.7) it is easily demonstrated that the WD also possesses the group delay property (12.24).

13. Linear signal transformations and their impact on ED's

In section 8 it has been shown that time and frequency shifts can very easily be observed in time–frequency shift invariant ED's, i.e. ED's that belong to the Cohen class. However, the delay of a signal by a certain time t_0 and its modulation with a carrier frequency f_0 constitute but the simplest signal processing operations. On the other hand, it has been observed in the foregoing section that it would be desirable to characterize not only signals but also linear time-invariant systems by ED's. In fact, the impulse response of a system is a special signal that of course can be represented by an ED. On the other hand, the same ED could also be used to characterize the signal processing operation of the system, i.e. the convolution in the time domain or the multiplication in the frequency domain. These considerations lead to the following compatibility requirements:

An ED is compatible with convolution in the time domain if the ED $T_{\tilde{x}}$ of two convolved signals x and h,

$$\tilde{x}(t) = \int_{t'} h(t - t')x(t') \, dt', \qquad (13.1)$$

can be obtained from T_x and T_h by

$$T_{\tilde{x}}(t, f) = \int_{t'} T_h(t - t', f) \, T_x(t', f) \, dt'. \tag{13.2}$$

Similarly, an ED is compatible with multiplication in the time domain if the ED $T_{\tilde{x}}$ of the two multiplied signals x and m,

$$\tilde{x}(t) = m(t) \cdot x(t), \tag{13.3}$$

can be obtained from T_x and T_m by

$$T_{\tilde{x}}(t, f) = \int_{f'} T_m(t, f - f') \, T_x(t, f') \, df'. \tag{13.4}$$

The corresponding constraint on the kernel function u_T is given by

$$u_T(t, f; t_1 + t_2, \tau_1 + \tau_2) = \int_{t'} u_T(t - t', f; t_1, \tau_1)$$
$$\times \, u_T(t', f; t_2, \tau_2) \, dt' \tag{13.5}$$

for the compatibility with convolution and the corresponding constraint on the kernel function U_T is given by

$$U_T(t, f; f_1 + f_2, \eta_1 + \eta_2) = \int_{f'} U_T(t, f - f'; f_1, \eta_1)$$
$$\times \, U_T(t, f'; f_2, \eta_2) \, df' \tag{13.6}$$

for the compatibility with multiplication in the time domain.

If we restrict the ED's considered to members of the shift-invariant class C_x, then the corresponding constraints on the ambiguity kernel function $\Psi_c(\tau, \eta)$ are given by

$$\Psi_c(\tau + \tau', \eta) = \Psi_c(\tau, \eta) \cdot \Psi_c(\tau', \eta) \tag{13.7}$$

for the (time) convolution compatibility, while for the compatibility with multiplication in the time domain, it is required that

$$\Psi_c(\tau, \eta + \eta') = \Psi_c(\tau, \eta) \cdot \Psi_c(\tau, \eta'). \tag{13.8}$$

14. Uniqueness of the Wigner distribution

In the foregoing sections many desirable properties have been motivated for ED's in view of their application in signal processing. In the present section it will be shown that the WD is unique in the sense that it is that single ED that possesses *all* of the stated desirable properties.

Let us consider the following set of properties:
- The bilinear signal representation should be time–frequency shift invariant. Then it follows from section 8 that it can be interpreted as a time–frequency weighted Wigner distribution as in eq. (8.13), but it can also be obtained from the ambiguity function as in eq. (8.16):

$$C_x(t, f) = \int_\tau \int_\eta \Psi_c(\tau, \eta) \, e^{j2\pi(t\eta - f\tau)} A_x(\tau, \eta) \, d\tau \, d\eta. \qquad (14.1)$$

- The bilinear signal representation should be an ED, see section 7. The average in time (7.2) is satisfied if the kernel $\Psi_c(\tau, \eta)$ fulfills the constraint

$$\Psi_c(0, \eta) = 1 \quad \text{for all } \eta. \qquad (14.2)$$

Similarly, the average in frequency (7.1) is satisfied if the kernel Ψ_c has the property

$$\Psi_c(\tau, 0) = 1 \quad \text{for all } \tau. \qquad (14.3)$$

- The normalized first-order moment in frequency of the ED should be given by the instantaneous frequency, see section 12. An ED has this property only then if its kernel Ψ_c meets the condition

$$\frac{\partial}{\partial \tau} \Psi_c(\tau, \eta) \bigg|_{\tau = 0} = 0 \quad \text{for all } \eta. \qquad (14.4)$$

A similar condition results if the ED should also have the group delay property (12.34):

$$\frac{\partial}{\partial \eta} \Psi_c(\tau, \eta) \bigg|_{\eta = 0} = 0 \quad \text{for all } \tau. \qquad (14.5)$$

- Imposing the compatibility of the ED with convolutions in the time and in the frequency domain. the conditions (13.7) and (13.8) on the kernel Ψ_c result in

$$\Psi_c(\tau + \tau', \eta) = \Psi_c(\tau, \eta) \cdot \Psi_c(\tau', \eta), \qquad (14.6)$$

$$\Psi_c(\tau, \eta + \eta') = \Psi_c(\tau, \eta) \cdot \Psi_c(\tau, \eta'). \qquad (14.7)$$

These last stated conditions on Ψ_c are both of the same form and are recognized as those functional relationships that can only be fulfilled by an exponential function. The general solution of (14.6) is therefore given by

$$\Psi_c(\tau, \eta) = e^{\alpha(\eta) \cdot \tau}, \qquad (14.8)$$

where $\alpha(\eta)$ is still an arbitrary function of η.

Yet, this Ψ_c has already property (14.2). If the instantaneous frequency should also be displayed from (14.4) this gives the condition

$$\left.\frac{\partial}{\partial\tau}\Psi_c(\tau,\eta)\right|_{\tau=0} = \alpha(\eta) = 0 \quad \text{for all } \eta. \tag{14.9}$$

From this, it follows that only

$$\Psi_c(\tau,\eta) = 1 \tag{14.10}$$

is the solution of (14.6) and (14.4).

But then (14.1) is just the 2-D Fourier transform of the ambiguity function, which is the WD:

$$C_x(t,f) = W_x(t,f) = \int_\tau \int_\eta A_x(\tau,\eta)\, e^{j2\pi(t\eta-f\tau)}\, d\tau\, d\eta. \tag{14.11}$$

It is easily seen that the other mentioned desirable properties are also met if $\Psi_c = 1$, i.e. the WD possesses them too.

15. Moyal's formula

In section 13 it occurred that the compatibility of certain signal processing operations with ED's required that corresponding ED-processing operations in the time–frequency plane hold, see (13.2) and (13.4).

A more general relation, originally found by Moyal (1949) for the WD, is

$$\int_t \int_f W_x(t,f) \cdot W_y(t,f)\, dt\, df = \left| \int_t x(t)\, y^*(t)\, dt \right|^2. \tag{15.1}$$

This can be interpreted as follows: the inner product of two WD's W_x and W_y in the time–frequency plane is always non-negative and is equal to the squared modulus of the inner product of the corresponding signals x and y.

Although it is possible to show (Hlawatsch 1985) that (15.1) holds for all ED's with unitary kernel functions $u_T(t,f; t_1, t_2)$ this idea will not be pursued in this course. Here only two special cases and consequences of (15.1) will be discussed.

The first case is obtained if the signal y is taken to be the delayed and modulated signal x. With the notation of section 8 this can be expressed by

$$y(t) = \mathcal{M}_{f_0}\, \mathcal{S}_{t_0}\, x(t) = x(t-t_0)\, e^{j2\pi f_0 t}. \tag{15.2}$$

The time delay and the frequency shift of the signal lead to the time delay and the frequency shift of its WD:

$$W_y(t, f) = W_x(t - t_0, f - f_0),$$ (15.3)

and the squared modulus of the inner product of the signals x and y is easily shown to be the squared modulus of the ambiguity function of $x(t)$:

$$|A_x(t_0, f_0)|^2 = \left| \int_t x(t) x^*(t - t_0) e^{-j2\pi f_0 t} dt \right|^2,$$ (15.4)

$$A_x(\tau, \eta) = \int_t x(t + \tfrac{1}{2}\tau) x^*(t - \tfrac{1}{2}\tau) e^{-j2\pi\eta t} dt,$$ (15.5)

$$= e^{j\pi\eta\tau} \int_t x(t) x^*(t - \tau) e^{-j2\pi\eta t} dt.$$ (15.6)

Combining eqs. (15.1)–(15.4) gives the result

$$\int_t \int_f W_x(t, f) W_x(t - t_0, f - f_0) \, dt \, df = |A_x(t_0, f_0)|^2,$$ (15.7)

i.e. the autocorrelation of the WD of the signal x in the (t, f)-plane is equal to the squared modulus of the corresponding ambiguity function. This result has very important consequences for the solution of detection problems.

Another special case of Moyal's formula deals with the structure of the WD itself. In section 5 it was observed that the bilinear nature of the considered signal representations leads to cross signal representations if the superposition of signals is considered. The cross WD of the signals x and y is given by

$$W_{xy}(t, f) = \int_\tau x(t + \tfrac{1}{2}\tau) y^*(t - \tfrac{1}{2}\tau) e^{-j2\pi f \tau} d\tau$$ (15.8)

and this can also be written as an inner product:

$$W_{xy}(t, f) = 2 \int_\tau \left(x(t + \tau) e^{-j2\pi f \tau} \right) \cdot \left(y(t - \tau) e^{j2\pi f \tau} \right)^* d\tau.$$ (15.9)

Both factors of the integrand are recognized as delayed and modulated signals, the second one is additionally reversed in time. Again Moyal's formula (15.1) can be applied, leading to the result

$$\int_\tau \int_\eta W_x\left(t + \tfrac{1}{2}\tau, f + \tfrac{1}{2}\eta\right) W_y\left(t - \tfrac{1}{2}\tau, f - \tfrac{1}{2}\eta\right) d\tau d\eta = \left|W_{xy}(t, f)\right|^2.$$

$$(15.10)$$

This relation very easily explains the structures and positions of inter-ference terms in Wigner distributions (Hlawatsch 1984).

16. Optimality of the WD with respect to energy concentration

In section 8 it was indicated that time–frequency shift invariance of ED's leads naturally to Cohen's class of bilinear ED's

$$C_x(t, f) = \int_\tau \int_\eta \psi_c(t - \tau, f - \eta) W_x(\tau, \eta) d\tau d\eta. \qquad (16.1)$$

$\psi_c(t, f)$ is the kernel function that characterizes the particular ED C_x as a smoothed WD. ψ_c fulfills the conditions (8.14) and (8.15). Because all bilinear ED's will attain negative values for certain specific signals x (see the discussion in section 7) a local and pointwise interpretation of C_x as an energy distribution function must be handled with care.

A result that very nicely reflects the energy concentration property of an ED and the time–frequency uncertainty imposed by Heisenberg's relation (10.6) has been found by Janssen (1984):

Each ED C_x of Cohen's class is smoothed according to

$$S_x(t, f) = \int_\tau \int_\eta g(t - \tau, f - \eta) C_x(\tau, \eta) d\tau d\eta \qquad (16.2)$$

with a Gaussian window function

$$g(t, f) = \exp\left[-2\pi\left(\frac{t^2}{T^2} + \frac{f^2}{F^2}\right)\right]. \qquad (16.3)$$

The parameters T and F can attain arbitrary positive values and their product TF determines the area in the (t, f) plane that is used for smoothing the C_x. Now three different cases have to be distinguished:

(1) $TF < 1$: The averaging area is "smaller" than Heisenberg's uncertainty ellipse. In this case there does not exist an ED C_x that yields non-negative values of S_x in (16.2) for all signals $x(t)$.

(2) $TF = 1$: The averaging area is just as large as Heisenberg's uncertainty ellipse. In this case the only member of Cohen's class that yields non-negative values of S_x for all signals $x(t)$ is the Wigner distribution.

(3) $TF > 1$: If the averaging area in (16.2) is larger than Heisenberg's uncertainty ellipse, then there exist several members of Cohen's class that yield non-negative values S_x for all signals $x(t)$.

In the sense of point (2) the Wigner distribution can be considered the most concentrated ED of Cohen's class. From the signal-theoretical point of view this statement might be more convincing than the more formal considerations in section 14 that the WD has superior signal representation qualities.

17. The discrete-time Wigner distribution

Until now the bilinear time–frequency signal representations of continuous-time analog signals were considered. However, for a practical application and the computation of the BSR's either special analog circuitry or digital signal processing methods have to be used.

Especially for the WD the case of discrete time signals has been investigated (Claasen and Mecklenbräuker 1980, 1983) and this will be discussed below. For a band-limited analog signal with cut-off frequency f_c the sampling theorem can be used to obtain a discrete time-signal representation:

$$x(t) = \sum_{n=-\infty}^{\infty} x(nT) \operatorname{sinc}\left(\frac{t}{T} - n\right), \tag{17.1}$$

where

$$\operatorname{sinc} t = \frac{\sin \pi t}{\pi t} \quad \text{and} \quad T = \frac{1}{f_s} \leqslant \frac{1}{2f_c}.$$

For these signals x the corresponding WD W_x will also be band-limited in frequency:

$$W_x(t, f) = 0 \quad \text{for} \quad |f| > f_c. \tag{17.2}$$

For $|f| < f_c$ it can be shown that W_x is given by

$$W_x(t, f) = \sum_{n=-\infty}^{\infty} h(t, f; n) \tilde{W}_x(nT, f), \tag{17.3}$$

where

$$\tilde{W}_x(nT, f) = \sum_{k=-\infty}^{\infty} x(kT) x^*([n-k]T) e^{-j2\pi f(2k-n)T} \tag{17.4}$$

and

$$h(t, f; n) = 2T(1 - 2T|f|) \operatorname{sinc}\left[\left(2\frac{t}{T} - n\right)(1 - 2|f|T)\right]. \tag{17.5}$$

While $\tilde{W}_x(nT, f)$ in (17.4) has the form of a discrete-time WD it can be shown that its discrete-time values are not the samples of the continuous-time WD (17.3). This means that for a good visualization of a WD the interpolation of W_x by the interpolation function h has to be performed as indicated in (17.3). However, this interpolation can be a heavy computational task, because a double sum has to be evaluated.

A true discrete-time WD $W_x(nT, f)$ is obtained, if instead of the minimum sampling frequency $f'_s = 2f_c$ in (17.1) the sampling frequency f_s is chosen to be higher than $f''_s = 4f_c$. Then the discrete-time signal $x(nT)$ is oversampled by a factor of at least two and the discrete-time WD is given by

$$W_x(nT, f) = 2 \sum_{k=-\infty}^{\infty} x([n+k]T)x^*([n-k]T)e^{-j4\pi kTf}. \quad (17.6)$$

In this form and with the above mentioned condition on the sampling period T the discrete-time values of $W_x(nT, f)$ of eq. (17.6) are identical with the samples of the continuous time Wigner distribution $W_x(t, f)$ of eq. (17.3).

18. The pseudo Wigner distribution

Although the time–frequency ED's are signal representations that unlike the Fourier transform give also an idea of the instantaneous signal behaviour their definition still depends on the total signal duration that can be rather long in general. However, for computational purposes it will be necessary to weight the signal x by a real-valued window function w of finite length before the corresponding ED is evaluated.

For the WD this idea leads to the pseudo Wigner distribution (Claasen and Mecklenbräuker 1980) which computes estimates of the WD from windowed signals:

$$x_t(\tau) = x(\tau)w(\tau - t), \quad (18.1)$$

$$W_{x_t}(t', f) = \int_\tau x_t(t' + \tfrac{1}{2}\tau) x_t^*(t' - \tfrac{1}{2}\tau) e^{-j2\pi f\tau} d\tau \quad (18.2)$$

$$= \int_\tau x(t' + \tfrac{1}{2}\tau)x^*(t' - \tfrac{1}{2}\tau)w(t' + \tfrac{1}{2}\tau - t)$$

$$\times w(t' - \tfrac{1}{2}\tau - t)e^{-j2\pi f\tau}d\tau. \quad (18.3)$$

If this WD is computed only for $t' = t$, i.e. for the center of the windowed signal, the pseudo WD is obtained that for each time instant t

is based on a different windowed signal $x_t(\tau)$:

$$\text{PWD}_x(t, f) = W_{x_t}(t', f)\big|_{t' = t} \tag{18.4}$$

$$= \int_\tau x\left(t + \tfrac{1}{2}\tau\right) x^*\left(t - \tfrac{1}{2}\tau\right) w\left(\tfrac{1}{2}\tau\right) w\left(-\tfrac{1}{2}\tau\right) e^{-j2\pi f\tau} d\tau. \tag{18.5}$$

This PWD can be shown to be a smoothed version of the true WD, where the smoothing is performed in the frequency direction only.

A similar idea has to be applied if the discrete-time WD in eq. (17.6) is to be computed. The discrete-time PWD is given by

$$\text{PWD}_x(nT, f) = 2 \sum_{k = -\infty}^{\infty} e^{-j4k\pi fT} x([n + k]T)\, x^*([n - k]T)$$
$$\times w(kT)\, w(-kT). \tag{18.6}$$

If the window has finite length then the PWD in eq. (18.6) can be computed by any DFT algorithm, which can be done efficiently using fast transform algorithms like the FFT.

All examples in section 20 are obtained by applying eq. (18.6) with an appropriate window function.

19. Several application areas

The roots of bilinear signal representations are certainly to be found in physics (Wigner 1932), while their importance for communications, signals and systems has been discovered much later (Koenig et al. 1946, Potter et al. 1947, Ville 1948).

However from 1979 on an ever increasing interest in this type of signal representation can be observed. Here only some research areas will be mentioned where BSR's have been investigated and applied:
- detection of signals (Woodward 1953, Altes 1980, Kay and Boudreaux-Bartels 1985);
- optical signals and systems (Bastiaans 1978–1981, 1983);
- seismic signal processing (Bouachache, Escudié and Komatitsch 1979);
- object recognition (Jacobson and Wechsler 1982);
- 2-D image processing (Jacobson and Wechsler 1982);
- description of nonstationary processes (Martin 1982);
- signal estimation, signal synthesis (Boudreaux-Bartels 1983);
- analysis of loudspeakers (Janse and Kaizer 1983);
- analysis of ultrasonic transducers (Marinovic and Smith 1984).

20. Examples of wigner distributions

The following abbreviations are used in the figure captions:

WD = Wigner distribution,   PWD = Pseudo Wigner distribution,
FM = frequency modulation,   AM = amplitude modulation.

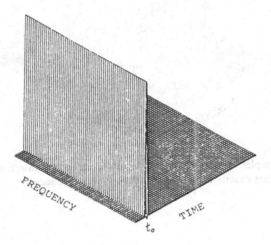

Fig. 1. WD of Dirac impulse at time t_0. Occurs exactly at t_0 and is constant wrt frequency.

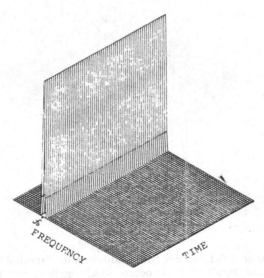

Fig. 2. WD of complex exponential with frequency f_0. Dual to fig. 1. "Spectral line" at f_0 which is constant wrt time.

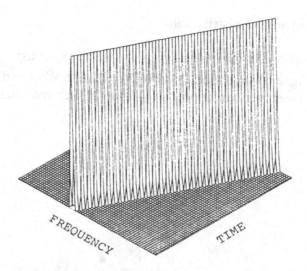

Fig. 3. WD of complex chirp signal, i.e. FM with linear modulation law. Generalization of figs. 1 and 2. Occurs exactly at the line of instantaneous frequency.

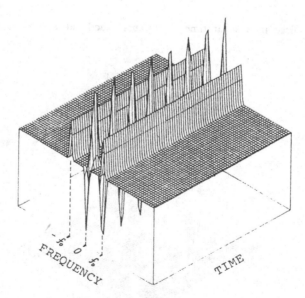

Fig. 4. WD of cosine with frequency f_0. In addition to the "spectral lines" at $\pm f_0$ (cf. fig. 2), an "interference term" at $f = 0$ occurs which oscillates wrt time with frequency $2f_0 = (+f_0 - (-f_0))$. This interference term reflects the oscillations of instantaneous power.

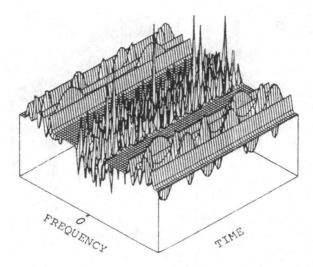

Fig. 5. WD of superposition of three cosine functions. There are interference terms between any two of the six spectral line terms.

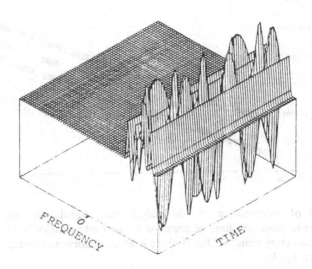

Fig. 6. WD of analytic signal corresponding to fig. 5. The band at negative frequencies and the "interference band" around $f = 0$ are suppressed.

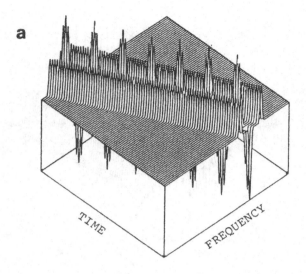

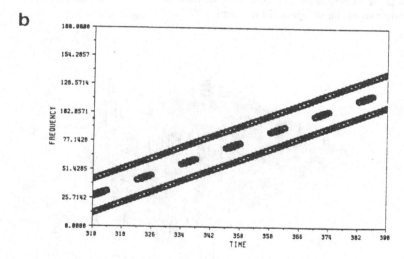

Fig. 7. (a) WD of superposition of two analytic chirp signals with parallel lines of instantaneous frequencies; (b) lines of constant height (only positive heights drawn). In addition to the two chirp terms (cf. fig. 3), there is an oscillatory interference term midway between them (cf. fig. 4).

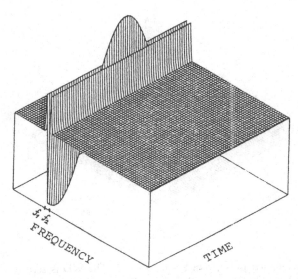

Fig. 8. WD of superposition of two cosine functions with nearly equal frequencies f_1, f_2 (analytic signal). The interference term occurs at the mid frequency $(f_1 + f_2)/2$ and oscillates with the difference frequency $f_2 - f_1$.

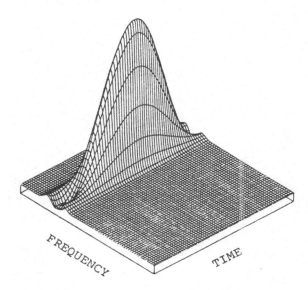

Fig. 9. Signal as in fig. 8, but PWD with sufficiently short window. PWD is a smoothed version of WD (with smoothing wrt frequency only). This causes the three distinct subterms of fig. 7 to merge into a single oscillating term which is now almost everywhere positive. The oscillation of this term with the difference frequency $f_2 - f_1$ clearly displays the beat between the two cosine functions (AM signal).

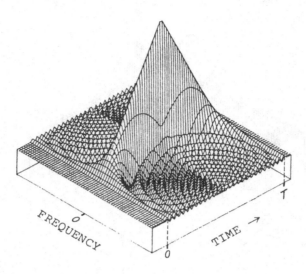

Fig. 10. WD of rectangular pulse from $t = 0$ to $t = T$. The WD is zero before $t = 0$ and after $t = T$ and is broad-band at $t = 0$ and $t = T$; towards the center point $t = T/2$ it sharpens into a "spectral line" at $f = 0$.

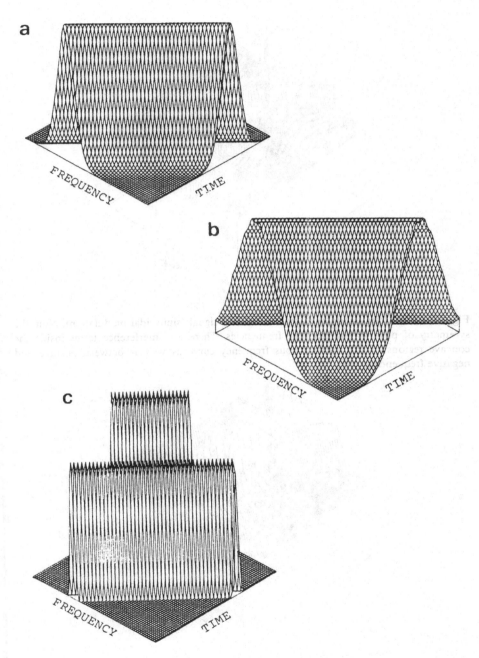

Fig. 11. Complex chirps signal. (a) Spectrogram with optimum window length (no better resolution can be obtained from the spectrogram). (b) Spectrogram with window too long. (c) PWD with window length as in (b) (additional term in the background due to aliasing). Note superior resolution of PWD.

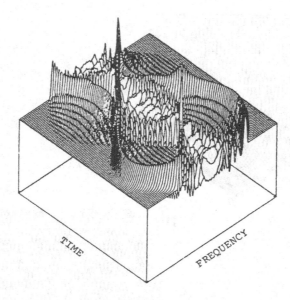

Fig. 12. PWD (long window) of real-valued FM signal (sinusoidal modulation). Note the symmetry of positive and negative frequencies. There are interference terms inside the concave regions of the instantaneous frequency curve as well as between positive and negative frequencies.

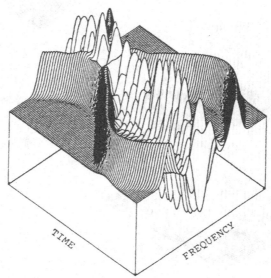

Fig. 13. PWD (short window), signal as in fig. 12. Due to the greater amount of smoothing, the interference terms inside the concave regions of the instantaneous frequency curve are cancelled, and frequency resolution is inferior.

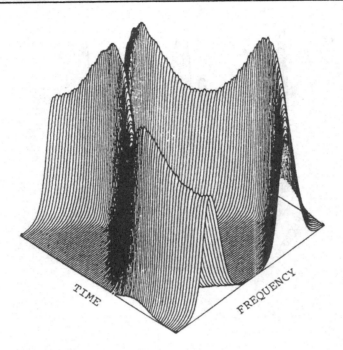

Fig. 14. Spectrogram, signal as in figs. 12 and 13, window length as in fig. 13. Interference terms are cancelled. The frequency resolution depends on the local slope of the instantaneous frequency curve (see fig. 15) and is generally inferior compared to PWD.

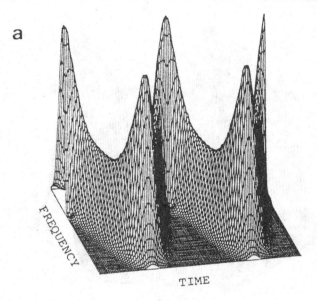

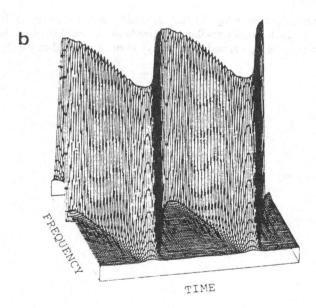

Fig. 15. Analytic FM signal (sinusoidal modulation): (a) Spectrogram, (b) PWD with same window length. The spectrogram is peaked if the instantaneous frequency is locally stationary but broad if the instantaneous frequency is locally chirp-like. The PWD, on the contrary, shows near-to-constant height and breadth at all time points (if the window is not too long).

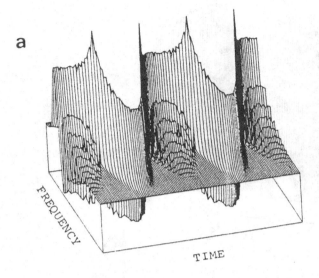

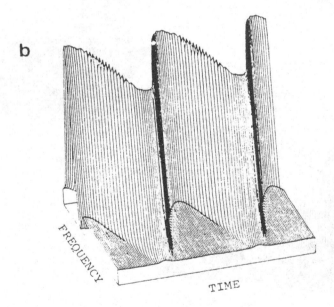

Fig. 16. PWD of analytic FM signal (sinusoidal modulation): (a) long window, (b) short window. To cancel interference terms (as in (b)), the window must be chosen short enough so that, at each time, the instantaneous frequency can be approximated over the window length by a straight line (local chirp approximation).

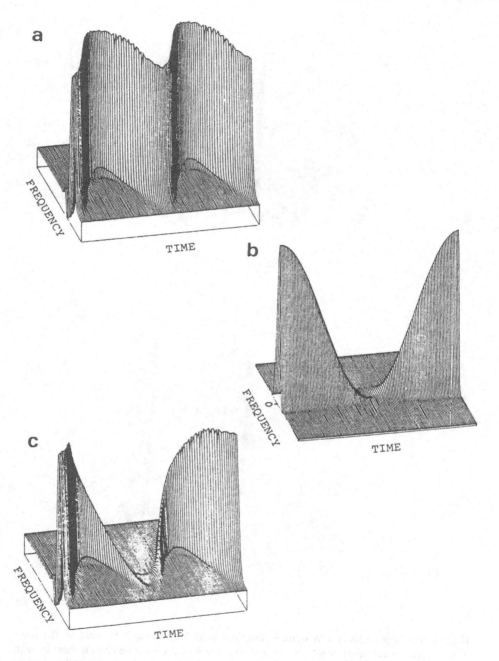

Fig. 17. PWD of (a) sinusoidal FM, (b) sinusoidal AM, (c) both. If the window is chosen properly (not too long), then FM and AM are displayed by PWD through separate features: instantaneous frequency-base curve; instantaneous power of envelope-height. This behaviour cannot be obtained from spectrograms (cf. fig. 14).

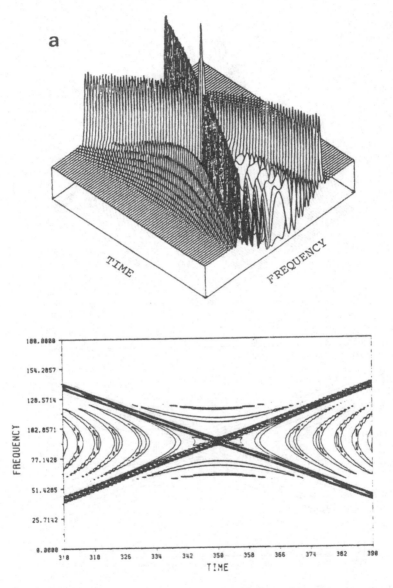

Fig. 18. Superposition of two complex chirp signals with lines of instantaneous frequencies intersecting. (a) PWD (long window), (b, overleaf) spectrogram (shorter window), (c, p. 321) spectrogram (longer window).

b

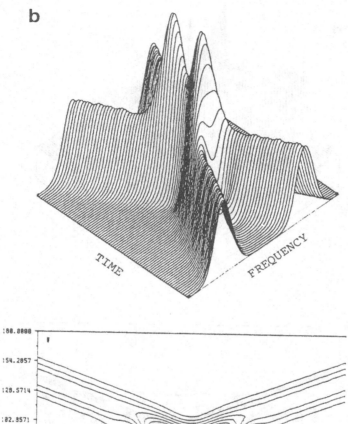

Fig. 18b.

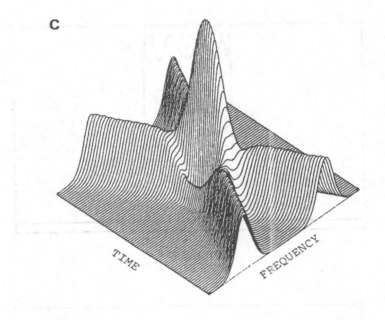

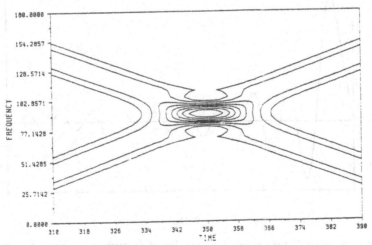

Fig. 18c.

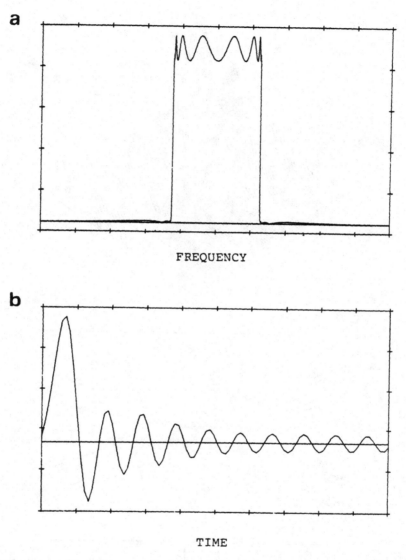

Fig. 19. Low-pass filter (Cauer type): (a) modulus of frequency response; (b) impulse response; (c) WD of impulse response. The long "tails" at the cut-off frequencies reflect the increase of group delay at these points.

C

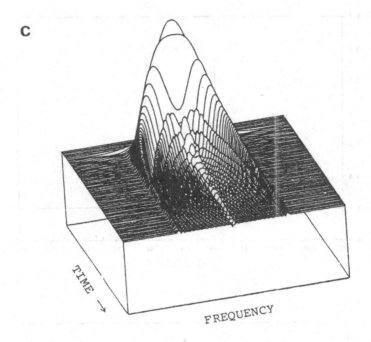

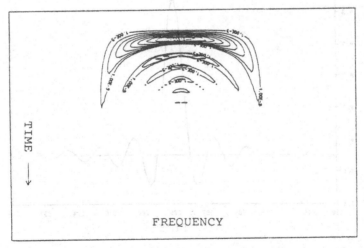

Fig. 19c.

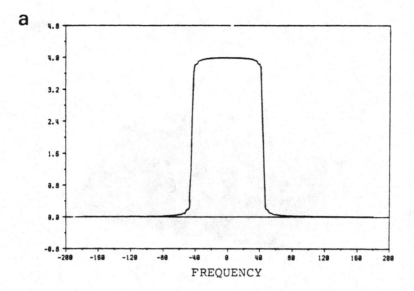

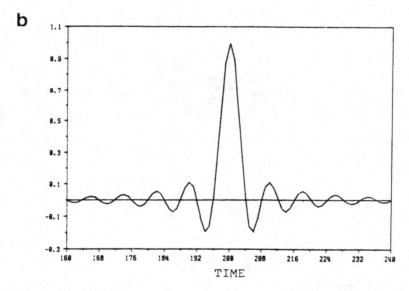

Fig. 20. As in fig. 19, but linear-phase low-pass. WD is symmetric wrt time, reflecting constant group delay. Nonetheless, there is still a broadening in time at the cut-off frequencies.

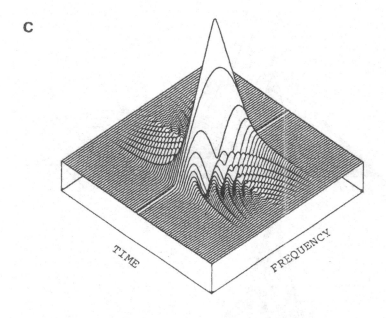

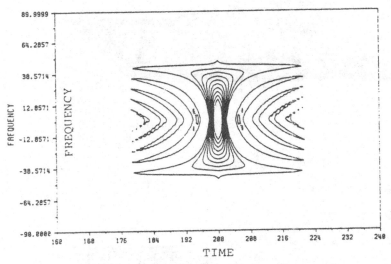

Fig. 20c.

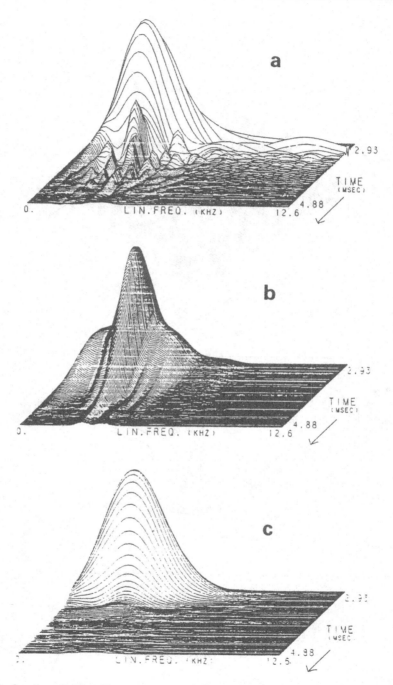

Fig. 21. Loudspeaker impulse response: (a) WD; (b) spectrogram with long window; (c) spectrogram with short window. All details are suppressed in the spectrogram: (b) is smeared in time, (c) is smeared in frequency. (From Janse and Kaizer 1983.)

Acknowledgement

The discussions with Mr. F. Hlawatsch and his support in generating the figures and the reference list are gratefully acknowledged.

References

In this reference list we have tried to give a comprehensive chronological compilation of the literature on bilinear signal representations and their applications.

1932

Wigner, E.P., On the quantum correction for thermo-dynamic equilibrium, Phys. Rev. **40**, 749–759.

1946

Gabor, D., Theory of communication, J. IEE (London) **93** (III) (November), 429–457.
Koenig, R., H.K. Dunn and L.Y. Lacy, The sound spectrograph, J. Acoust. Soc. Amer. **18**, 19–49.

1947

Potter, R.K., G.A. Kopp and H.C. Green, Visible Speech (Van Nostrand, New York). Republished by Dover Publications, 1966.

1948

Ville, J., Théorie et applications de la notion de signal analytique, Câbles et Transmission **2A**(1), 61–74 [Engl. Transl.: I. Selin, Rand Corp. Report T-92, August 1, 1958].

1949

Moyal, J.E., Quantum mechanics as a statistical theory, Proc. Camb. Phil. Soc. **45** (January), 99–124.

1950

Fano, R.M., Short-time autocorrelation functions and power spectra, J. Acoust. Soc. Amer. **22**(5) (September), 546–550.

1952

Page, C.H., Instantaneous power spectra, J. Appl. Phys. **23**, 103–106.

1953

Woodward, P.M., Probability and information theory with application to radar (Pergamon Press, London).

1955

Blanc-Lapierre, A., and B. Picinbono, Remarques sur la notion du spectre instantané de puissance, Publ. Sci. Univ. d'Alger, Série B1, 2–32.

1962

More, H., I. Oppenheim and J. Ross, Some topics in quantum statistics: the Wigner function and transport theory, in: Studies in Statistical Mechanics, ed. G.E. Uhlenbeck, (North-Holland Publishing, Amsterdam) pp. 213–298.

Schroeder, M.R., and B.S. Atal, Generalized short-time power spectra and auto-correlation function, J. Acoust. Soc. Amer. **34**(11), (November), 1679–1683.

1964

Levin, M.J., Instantaneous power spectra and ambiguity function, IEEE Trans. Inf. Theory **IT-10**, 95–97.

1966

Cohen, L., Generalized phase-space distribution functions, J. Math. Phys. **7**, 781–786.

Helström, C.W., An expansion of a signal into Gaussian elementary signals, IEE Trans. Inf. Theory **IT-12**, 81–82.

1967

DeBruijn, N.G., Uncertainty principles in Fourier analysis, in: Inequalities, ed. O. Shisha (Academic Press, New York) pp. 57–71.

Levin, M.J., Instantaneous spectra and ambiguity functions, IEEE Trans. Inf. Theory **IT-13**, 95–97.

Margenau, H., and L. Cohen, Probabilities in quantum mechanics, in: Quantum theory and reality, ed. M. Bunge (Springer, Berlin) ch. 4, pp. 71–89.

Montgomery, L.K., and I.S. Reed, A generalization of the Gabor–Helström transform, IEEE Trans. Inf. Theory **IT-13**, 344–345.

1968

Bonnet, G., Considérations sur la représentation et l'analyse harmonique des signaux déterministes ou aléatoires, Ann. Télécom. **23** (3–4), 62–86.

Rihaczek, A.W., Signal energy distribution in time and frequency, IEEE Trans. Inf. Theory **IT-14**, 369–374.

1970

Ackroyd, M.H., Short-time spectra and time–frequency energy distributions, J. Acoust. Soc. Amer. **50**(5), 1229–1231.

Ackroyd, M.H., Instantaneous and time-varying spectra – An introduction, Radio & Electron. Eng. **39**, 145–152.

Ackroyd, M.H., Instantaneous spectra and instantaneous frequency, Proc. IEEE **58**(1), p. 141.

Mark, W.D., Spectral analysis of the convolution and filtering of non-stationary stochastic processes, J. Sound & Vib. **11**(1), 19–63.

Oppenheim, A.V., Speech spectrograms using the fast Fourier transform, IEEE Spectrum **7** (August), 57–62.

1971

Ruggeri, G.J., On phase-space description of quantum mechanics, Prog. Theor. Phys. **46**(6), 1703–1712.

1972

Escudié, B., Représentation temps–fréquence dans l'analyse et la synthèse des signaux. in: 6è Congrès International de Cybernétique, pp. 277–299.

Levshin, A., V.F. Pisarenko and G.A. Pogrebinsky, On a frequency–time analysis of oscillations, Ann. Geophys. **28**, 211–218.

Nelson, G.A., Signal analysis in time and frequency using Gaussian wavefunctions. in: NATO Advanced Studies Institute on Network and Signal Theory, ed. J.O. Scalan (September 1972) pp. 454–460.

1973

DeBruijn, N.G., A theory of generalized functions with applications to Wigner distribution and Weyl correspondence, Nieuw Arch. Wiskunde **21**, 205–280.

1975

Lacoume, J.L., and W. Kofman. Description des processus non-stationnaires par la représentation temps–fréquence. 5è Coll. Traitement du Signal et applications GRETSI, 14/1–14/7 (Nice).

1976

Escudié, B., and J. Gréa, Sur une formulation générale de la représentation en temps et fréquence dans l'analyse de signaux d'énergie finie, C. R. Acad. Sci. Paris, série A **283**, 1049–1051.

Kodera, K., C. DeVilledary and R. Gendrin. A new method for the numerical analysis of non-stationary signals, Phys. Earth & Planet. Inter. **12**, 142–150.

Krüger, J.G., and A. Poffyn, Quantum mechanics in phase space. Part I – Unicity of the Wigner distribution function. Physica **85A**, 84–100.

Tjostheim, D., Spectral generating operators for non-stationary processes. Adv. Appl. Prob. **8**, 831–846.

1977

Allen, J.B., and L.R. Rabiner, A unified approach to short-time Fourier analysis and synthesis, Proc. IEEE **65**, 1558–1564.

Berry, M.V., Semi-classical mechanics in phase space: a study of Wigner's function, Phil. Trans. Roy. Soc. A **287**, 237.

Escudié, B., and J. Gréa, Représentation Hilbertienne et représentation conjointe en temps et fréquence des signaux d'énergie finie, Coll. GRETSI, Nice, France (April 1977).

Royer, A., Wigner distribution as the expectation value of a parity operator, Phys. Rev. A**15**(2), 449, 450.

1978

Bastiaans, M.J., The Wigner distribution function applied to optical signal and systems, Opt. Commun. **25**, 26–30.

Kodera, K., R. Gendrin and C. DeVilledary, Analysis of time-varying signals with small BT values, IEEE Trans. Acoust., Speech & Signal Process. **ASSP-26**(1), 64–76.

Melard, G., Propriétés du spectre évolutif d'un processus non-stationnaire. Ann. Inst. Henri Poincaré B **XIV**(4), 411–424.

1979

Bastiaans, M.J., Wigner distribution function and its application to first order optics, J. Opt. Soc. Amer. **69**, 1710–1716.

Bastiaans, M.J., The Wigner distribution function and Hamiltonian's characteristics of a geometrical–optical system, Opt. Commun. **30**(3), 321–326.

Bastiaans, M.J., Transport equations for the Wigner distribution function, Opt. Acta **26**, 1256–1272.

Bastiaans, M.J., Transport equations for the Wigner distribution function in inhomogeneous and dispersive media, Opt. Acta **26**, 1334–1344.

Bouachache, B., B. Escudié, P. Flandrin and J. Gréa, Sur une condition nécessaire et suffisante de positivité de la représentation conjointe en temps et fréquence des signaux d'énergie finie, C. R. Acad. Sci. Paris, Série A **288**, 307–309.

Bouachache, B., B. Escudié and J.M. Komatitsch, Sur la possibilité d'utiliser la représentation conjointe en temps et fréquence dans l'analyse des signaux modulés en fréquence émis en vibrosismique, 7ème Coll. GRETSI sur le Traitement du Signal et Applications, Nice, France (1979) pp. 121/1–121/6.

Bremmer, H., The Wigner distribution and transport equations in radiation problems, J. Appl. Soc. Eng. A **3**, 251–260.

Escudié, B., "Représentation en temps et fréquence des signaux d'énergie finie: analyse et observation des signaux, Ann. Télécommun. **34**(3–4), 101–111.

Escudié, B., P. Flandrin and J. Gréa, Positivité des représentations en temps et fréquence des signaux d'énergie finie, représentation Hilbertienne et conditions d'observation des signaux, 7ème Coll. GRETSI sur le Traitement du Signal et Applications, Nice, France (1979) pp. 28/5–2/6/79.

Fargetton, H., Fréquences instantanées de signaux multicomposantes, Thèse (Grenoble).

Gendrin, R., and C. DeVilledary, Unambiguous determination of fine structures in multicomponent time-varying signals, Ann. Télécommun. **35**(3–4), 122–130.

Janssen, A.J.E.M., Application of Wigner distributions to harmonic analysis of generalized stochastic processes, MC-tract 114 (Mathematisch Centrum, Amsterdam).

Rabiner, L.R., and J.B. Allen, Short-time Fourier analysis techniques for FIR system identification and power spectrum estimation, IEEE Trans. Acoust., Speech & Signal Process. **ASSP-27**.

Wigner, D.P., Quantum-mechanical distribution functions revisited, in: Perspectives in quantum theory, eds. W. Yourgran and A. van de Merwe (Dover, New York).

1980

Altes, R.A., Detection, estimation and classification with spectrograms. J. Acoust. Soc. Am. **67**(4), 1232–1246.

Bastiaans, M.J., Gabor's expansion of the signal into Gaussian elementary signals, Proc. IEEE **68**(4) p. 538.

Bastiaans, M.J., Wigner distribution function display: a supplement to ambiguity display using a single 1-D Input, Appl. Opt. **19**(2), 192.

Bartelt, H.O., K-H. Brenner and A.W. Lohmann, The Wigner distribution function and its optical production, Opt. Commun. **32**, 32–38.

Claasen, T.A.C.M., and W.F.G. Mecklenbräuker, The Wigner distribution – a tool for time–frequency signal analysis – Part I: Continuous-time signals, Philips J. Res. **35**(3), 217–250.

Claasen, T.A.C.M., and W.F.G. Mecklenbräuker, The Wigner distribution – a tool for time–frequency signal analysis – Part II: Discrete-time signals, Philips J. Res. **35** (4/5), 276–300.

Claasen, T.A.C.M., and W.F.G. Mecklenbräuker, The Wigner distribution – a tool for time–frequency signal analysis – Part III: Relations with other time–frequency signal transformations, Philips J. Res. **35**(6), 372–389.

Escudié, B., and P. Flandrin, Sur quelques propriétés de la représentation conjointe et de la fonction d'ambiguité des signaux d'énergie finie, C. R. Acad. Sci. Paris A **291**, 171–174.

Flandrin, P., Application de la théorie des catastrophes a l'étude du comportement asymptotique de la représentation conjointe de Wigner–Ville, DEA, CEPHAG, Grenoble, 1980.

Flandrin, P., and B. Escudié, Time and frequency representation of finite energy signals: a physical property as a result of a Hilbertian condition, Signal Process. **2**, 93–100.

Marcuvitz, N., Quasiparticle view of wave propagation, Proc. IEEE, **68**, pp. 1380–1395.

Portnoff, M.R., Time–Frequency Representation of Digital Signals and Systems based on Short-time Fourier Analysis, IEEE Trans. Acoust., Speech & Signal Process. **ASSP-28**(1), pp. 55–69.

1981

Bastiaans, M.J., The Wigner distribution function and its applications to optics, in: Optics in Four Dimensions, ed. L.M. Narducci (American Institute of Physics, New York) pp. 292–312.

Bastiaans, M.J., A sampling theorem for the complex spectrogram and Gabor expansion of a signal in Gaussian elementary signals, Opt. Eng. **20**(4), 594–598; in: Proc. Int. Opt. Computing Conf., Washington, D.C., 8–11 April, 1980.

Bertrand, P., and C. Fugier-Garrel, Formulation de la théorie de la communication dans le plan temps–frequence–aspects practiques," 8ème Coll. GRETSI, Nice (1981), pp. 829–834.

Claasen, T.A.C.M., and W.F.G. Mecklenbräuker, Time-frequency signal analysis by means of the Wigner distribution, Proc. IEEE Int. Conf. on Acoustics, Speech and Signal Processing, Atlanta, CA (1981), pp. 69–72.

Fargetton, H., R. Gendrin and J.L. Lacoume, Adaptive methods for spectral analysis of time-varying signals, in: Signal Processing: Theories and Applications–II, (Proc. EUSIPCO-80 Conf., Lausanne, September 1980) eds. M. Kunt and F. De Coulon (North-Holland, Amsterdam) pp. 787–792.

Flandrin, P., and B. Escudié, "Géometrie des fonctions d'ambiguité et des représentations conjointes de Ville: l'approche de la théorie des catastrophes, 8ème Coll. GRETSI, pp. 69–74, Nice (1981).

Flandrin, P., B. Escudié, J. Gréa and Y. Biraud, Joint representation and Hilbertian analysis: "elementary" and "asymptotic" finite energy signals, Proc. EUSIPCO-80 Conf., Lausanne (1980), eds. M. Kunt and F. de Coulon (North-Holland, Amsterdam) pp. 25–26.

Grace, O.D., Instantaneous power spectra, J. Acoust. Soc. Amer. **69**(1), 191–198.

Janssen, A.J.E.M., Positivity of weighted Wigner distributions, SIAM J. Math. Anal. **12**(5), 752–758.

Priestly, M.B., Spectral analysis and time series. Vols. I and II (Academic Press, New York).

Szu, H.H., and J.A. Blodgett, Wigner distribution and ambiguity function. in: Optics in Four Dimensions, ed. L.M. Narducci (American Institute of Physics, New York) pp. 355–381.

1982

Bouachache, B., Représentation temps–fréquence–Application à la mesure de l'absorption du sous-sol, Thèse D.I. (INPG, Grenoble).

Bouachache, B., and P. Flandrin, Wigner–Ville analysis of time-varying signals. Proc. IEEE Int. Conf. on Acoustics, Speech and Signal Processing, ICASSP82, Paris, France (May 1982) pp. 1329–1333.

Chan, D.S.K., A non-aliased discrete-time Wigner distribution for time–frequency signal analysis, Proc. IEEE Int. Conf. on Acoustics, Speech and Signal Processing. ICASSP82. Paris, France (May 1982) pp. 1333–1336.

Chester, D.B., The Wigner distribution and its application to speech recognition and analysis, Ph.D. Dissertation (University of Cincinnati, OH).

Flandrin, P., Représentation des signaux dans le plan temps–fréquence. Thèse D.I. (Grenoble).

Flandrin, P., and B. Escudié, Sur la localisation des représentations conjointes dans le plan temps–fréquence, C. R. Acad. Sci. Paris, série I **295**(7), 475–478, presented by A. Blanc-Lapierre.

Flandrin, P., B. Escudié and J. Gréa, Applications of operators and deformation techniques to time–frequency problems, IEEE ISIT, Les Arcs, 1982.

Jacobson, L., and H. Wechsler, A paradigm for invariant object recognition of brightness. optical flow and binocular disparity images, Pattern Recognition Lett. **1**, 61–68.

Jacobson, L., and H. Wechsler, The Wigner distribution and its usefulness for 2-D image processing, Sixth Int. Joint Conference on Pattern Recognition, Munich, FRG (October 19–22, 1982).

Janssen, A.J.E.M., On the locus and spread of pseudo-density functions in the time–frequency plane, Philips J. Res. **37**, 79–110.

Martin, W., Time–frequency analysis of non-stationary processes, IEEE-ISIT, Les Arcs, 1982.

Martin, W., Time–frequency analysis of random signals, Proc. IEEE Int. Conf. on Acoustics, Speech and Signal Processing, ICASSP82, Paris, France (May 1982) pp. 1325–1328.

Nawab, S.H., T.F. Quatieri and J.S. Lim, Signal reconstruction from short-time Fourier transform magnitude, Proc. IEEE Int. Conf. on Acoustics, Speech and Signal Processing ICASSP82, Paris, France (May 1982) pp. 1046–1048.

Szu, H.H. Two-dimensional optical processing of one-dimensional acoustic data, Opt. Eng. **21**, 804–813.

1983

Bastiaans, M.J., Signal description by means of a local frequency spectrum, doctoral thesis (Technical University, Eindhoven, The Netherlands).

Bouachache, B., Wigner–Ville analysis of time-varying signals: an application in seismic prospecting, EUSIPCO-83: Second European Association for Signal Processing Conference, Erlangen, FRG (September 12–16, 1983).

Boudreaux-Bartels, G.F., Time–frequency signal processing algorithms: analysis and synthesis using Wigner distributions, Ph.D. Dissertation (Rice University, Houston, TX).

Boudreaux-Bartels, G.F., and T.W. Parks, Reducing aliasing in the Wigner distribution using implicit spline interpolation, Proc. IEEE Int. Conf. on Acoustics, Speech and Signal Processing ICASSP83, Boston, MA (April 1983) pp. 1438–1441.

Brenner, K.H., A discrete version of the Wigner distribution function, EUSIPCO-83, Second European Association for Signal Processing Conference, Erlangen, FRG (September 12–16, 1983).

Chester, D., F. Taylor and M. Doyle, On the Wigner distribution, Proc. IEEE Int. Conf. on Acoustics, Speech and Signal Processing ICASSP83, Boston, MA (April 1983) Vol. 2, pp. 491–494.

Chester, D., F.J. Taylor and M. Doyle, Application of the Wigner distribution to speech processing, IEEE Acoustics, Speech and Signal Processing, Spectrum Estimation Workshop II, Tampa, Florida (November 10–11, 1983) pp. 98–102.

Claasen, T.A.C.M., and W.F.G. Mecklenbräuker, The aliasing problem in discrete-time Wigner distributions, IEEE Trans. Acoust., Speech & Signal Process. **ASSP-31**, 1067–1072.

Flandrin, P., and W. Martin, Sur les conditions physiques assurant l'unicité de la représentation de Wigner–Ville comme représentation temps–fréquence, Neuvième Colloque sur la Traitement du Signal et ses Applications, pp. 43–49 (May 1983).

Flandrin, P., and W. Martin, Pseudo-Wigner estimators, IEEE ASSP Spectrum Estimation Workshop II, Tampa, Florida (November 10–11, 1983) pp. 181–185.

Grenier, Y., Time-dependent ARMA modeling of non-stationary signals, IEEE Trans. Acoust. Speech & Signal Process. **ASSP-31**, pp. 899–911.

Grenier, Y., and D. Aboutajdine, Comparison des représentations temps–fréquence de signaux présentant des discontinuités spectrales, Ann. Telecom. **38**(11–12), 429–442.

Griffin, D.W. and J.S., Lim, Signal estimation from modified short-time Fourier transform, Proc. IEEE Int. Conf. on Acoustics, Speech and Signal Processing ICASSP83, Boston, MA (April 1983) pp. 804–807.

Jacobson, L., and H. Wechsler, The composite pseudo Wigner distribution (CPWD): A computable and versatile approximation to the Wigner distribution (WD), Proc. IEEE Int. Conf. on Acoustics, Speech and Signal Processing ICASSP83, Boston, MA (April 1983) pp. 254–256.

Janse, C.P. and A.J.M. Kaizer, Time–frequency distributions of loudspeakers: the application of the Wigner distribution, J. Audio Eng. Soc. **31**(4), 198–223.

Kraus, G., Eine Systematik der linearen, bilinearen und quadratischen Analyse deterministischer Signale, Arch. Elektron. Uebertragungstech. 37, Heft 5/6, 160–164.

Martin, W., and P. Flandrin, Analysis of non-stationary processes: short-time periodograms versus a pseudo-Wigner estimator, EUSIPCO'83, (North-Holland, Amsterdam).

Nawab, S.H., T.F. Quatieri and J.S. Lim, Algorithms for signal reconstruction from short-time Fourier transform magnitude, Proc. IEEE Int. Conf. on Acoustics, Speech and Signal Processing ICASSP83, Boston, MA (April 1983) pp. 800–803.

O'Connell, R.F., The Wigner distribution function – 50th birthday, Found. Phys. **13**(1), 83–92.

Taylor, F., M. Doyle and D. Chester, On the Wigner distribution, Proc. IEEE Int. Conf. on Acoustics, Speech and Signal Processing ICASSP83, Boston, MA (April 1983) pp. 491–494.

1984

Auslander, L., and R. Rolimieri, Characterizing the radar ambiguity functions, IEEE Trans. Inf. Theory **IT-30**(6), 832–836.

Bouachache, B., and F. Rodriguez, Recognition of time-varying signals in the time–frequency domain by means of the Wigner distribution, Proc. IEEE Int. Conf. on Acoustics, Speech and Signal Processing ICASSP84, San Diego, CA (March 1984) pp. 22.5.1–22.5.4.

Boudreaux-Bartels, G.F., and T.W. Parks, Signal estimation using modified Wigner distributions, Proc. IEEE Int. Conf. on Acoustics, Speech and Signal Processing ICASSP84, San Diego, CA (March 1984).

Brud, B.R., and T.E. Posch, A range and azimuth estimator based on forming the spatial Wigner distribution, Proc. IEEE Int. Conf. on Acoustics, Speech and Signal Processing ICASSP84, San Diego, CA (March 1984) pp. 41.B.9.1–41.B.9.2.

Chester, D., F.J. Taylor and M. Doyle, The Wigner distribution in speech processing applications, J. Franklin Inst. **318**(6), 415–430.

Claasen, T.A.C.M., and W.F.G. Mecklenbräuker, On the time–frequency discrimination of energy distributions: can they look sharper than Heisenberg?, Proc. IEEE Int. Conf. on Acoustics, Speech and Signal Processing ICASSP84, San Diego, CA (March 1984) pp. 41.B.7.1–41.B.7.4.

Cohen, L., Distributions in signal theory, Proc. IEEE Int. Conf. on Acoustics, Speech and Signal Processing ICASSP84, San Diego, CA (March 1984) pp. 41.B.1.1–41.B.1.4.

Escudié, B., and J. Gréa, Joint representation (JR) in signal theory (ST) and Hilbertian analysis: a powerful tool for signal analysis, Proc. IEEE Int. Conf. on Acoustics, Speech and Signal Processing ICASSP84, San Diego, CA (March 1984) pp. 41.B.6.1–41.B.6.4.

Flandrin, P., Some features of time–frequency representations of multicomponent signals, IEEE ICASSP84, San Diego (1984) pp. 41.B.4.1–41.B.4.4.

Flandrin, P., and B. Escudié, An interpretation of the pseudo-Wigner–Ville distribution, Signal Process 6(1), 27–36.

Flandrin, P., and W. Martin, A general class of estimators for the Wigner–Ville spectrum of non-stationary processes, in: Lectures Notes in Control and Information Sciences, Vol. 62, Analysis and Optimization of Systems (Springer-Verlag, Heidelberg) pp. 15–23.

Flandrin, P., W. Martin and M. Zakharia, On a hardware implementation of the Wigner–Ville transform, in: Digital Signal Processing-84, eds. V. Cappellini and A.G. Constantinides (North-Holland, Amsterdam) pp. 262–266.

Grenier, Y., Modélisation de signaux non-stationnaires, Thèse (Université de Paris-Sud).

Hillery, M., R.F. O'Connell, M.O. Scully and E.P. Wigner, Distribution functions in physics: fundamentals, Phys. Rep. 106(3), 123–167.

Hlawatsch, F., PseudoWignerverteilung modulierter Schwingungen, Fünfter Aachener Kolloquium, (Aachen, 1984) S. 344–347.

Hlawatsch, F., Interference terms in the Wigner distribution, in: Digital Signal Processing–84, eds. V. Cappellini and A.G. Constantinides (North-Holland, Amsterdam) pp. 363–367.

Janssen, A.J.E.M., Positivity properties of phase–space distribution function, J. Math. Phys. 25(7), 2240–2252.

Janssen, A.J.E.M., A note on Hudson's theorem about functions with nonnegative Wigner distributions, SIAM J. Math. Anal. 15(1), 170–176.

Janssen, A.J.E.M., Gabor representations and Wigner distribution of signals, Proc. IEEE Int. Conf. on Acoustics, Speech and Signal Processing ICASSP84, San Diego, CA (1984) pp. 41.B.2.1–41.B.2.4.

Kumar, B.V.K.V., and C.W. Carroll, Performance of Wigner distribution function based detection methods, Opt. Eng. 23(6), 732–737.

Marinovic, N.M., and W.A. Smith, Use of the Wigner distribution to analyse the time–frequency response of ultrasonic transducers, IEEE Ultrasonics Symposium, Dallas, TX (1984).

Martin, N., Développement de méthodes d'analyse spectrale autoregressive, Applications a des signaux réels non-stationnaires ou a N dimensions, Thèse D.I. (INPG, Grenoble).

Martin, W., Measuring the degree of non-stationarity by using the Wigner–Ville spectrum, Proc. IEEE Int. Conf. on Acoustics, Speech and Signal Processing ICASSP84, San Diego, CA (1984) pp. 41.B.3.1–41.B.3.4.

Martin, W., Spectral analysis of nonstationary processes, Sixth Int. Conf. on Analysis and Optimisation of Systems (special session on nonstationary processes), Nice, France (June 1984).

Martin, W., Wigner–Ville-Spektralanalyse nichtstationärer Prozesse, Theorie und Anwendung in biologischen Fragestellungen, Habilitationsschrift (Universität Bonn).

Riley, M.D., Detecting time-varying spectral energy concentrations, IEEE Digital Signal Processing Workshop, Chatham (1984) pp. 5.6.1–5.6.2.

Schempp, W., Radar ambiguity functions, nilpotent harmonic analysis, and holomorphic theta series, Special Functions: Group Theoretical Aspects and Applications, eds. R.A. Askey et al., pp. 217–260.

Subotic, N., and B.E.A. Salek, Generation of the Wigner distribution function of two-dimensional signals by a parallel optical processor, Opt. Lett. 9(10), 471–473.

Szu, H.H., and H.J. Caulfield, The mutual time–frequency content of two signals, Proc. IEEE 72(7), 902–908.

1985

Boashash, B., On the anti-aliasing and computational properties of the Wigner–Ville distribution, Int. Symp. on Applied Signal Processing and Digital Filtering IASTED85, Paris, (June 1985) ed. M.H. Hamza (Acta Press, Anaheim).

Chester, D., and J. Wilbur, Time and spatial varying CAM and AI signal analysis using the Wigner distribution, Proc. IEEE Int. Conf. on Acoustics, Speech and Signal Processing ICASSP85, Tampa. FL (1985) pp. 1045–1048.

Cohen, L., Properties of the positive time–frequency distribution functions, Proc. IEEE Int. Conf. on Acoustics, Speech and Signal Processing, ICASSP85, Tampa, FL (1985) pp. 548–551.

Cohen, L., Positive and negative-joint quantum distributions, in: Non-equilibrium Quantum Statistical Physics, eds. J. Moore and M.O. Scully, 1985.

Cohen, L., and T. Posch, Positive time–frequency distribution functions, IEEE Trans. Acoust. Speech & Signal Process. ASSP-33, 31–38.

Flandrin, P., Séparation de fréquences modulées proches par analyse de Wigner–Ville autoregressive, 10è Coll. Traitement du Signal GRETSI, Nice, France (May 1985).

Flandrin, P., and B. Escudié, Principe et mise en oeuvre de l'analyse temps–fréquence par transformation de Wigner–Ville, Traitement du Signal, GRETSI, Nice, France.

Flandrin, P., B. Escudié and W. Martin, Représentations temps–fréquence et causalité, Coll. GRETSI-85, Nice, France (May 1985).

Friedman, D.H., Instantaneous-frequency distribution vs. time: an interpretation of the phase structure of speech, Proc. IEEE Int. Conf. on Acoustics, Speech and Signal Processing ICASSP85, Tampa, FL (1985) pp. 1121–1124.

Garudadri, H., M.P. Beddoes, J.H.V. Gilbert and A.-P. Benguered, Identification of invariant acoustic cues in stop consonants using the Wigner distribution, Int. Symp. on Applied Signal Processing and Digital Filtering IASTED85, Paris (June 1985) ed. M.H. Hamza (Acta Press, Anaheim) pp. 196–200.

Hammond, J., and R. Harrison, Wigner–Ville and evolutionary spectra for covariant equivalent nonstationary random processes, Proc. IEEE Int. Conf. on Acoustics, Speech and Signal Processing ICASSP85, Tampa, FL (1985) pp. 1025–1028.

Hlawatsch, F., Transformation, inversion and conversion of bilinear signal representations, Proc. IEEE Int. Conf. on Acoustics, Speech and Signal Processing ICASSP85, Tampa, FL (1985) pp. 1029–1032.

Hlawatsch, F., Duality of time–frequency signal representations: energy density domain and correlation domain, Int. Symp. on Applied Signal Processing and Digital Filtering, IASTED85, Paris (June 1985) ed. M.H. Hamza (Acta Press, Anaheim).

Hlawatsch, F., Notes on bilinear time–frequency signal representations, Institutsbericht (Institut für Nachrichten-technik, Technische Universität Wien).

Kay, S., and G. Boudreaux-Bartels, On the optimality of the Wigner distribution for detection, Proc. IEEE Int. Conf. on Acoustics, Speech and Signal Processing ICASSP85, Tampa, FL (1985) pp. 1017–1020.

Marinovic, N.M., and G. Eichmann, An expansion of Wigner distribution and its applications, Proc. IEEE Int. Conf. on Acoustics, Speech and Signal Processing ICASSP85, Tampa, FL (1985) pp. 1021–1024.

Martin, W., and P. Flandrin, Detection of changes of signal structure by using the Wigner–Ville spectrum, Signal Process. 8, 215–233.

Yu, K., and S. Cheng, Signal synthesis from Wigner distribution, Proc. IEEE Int. Conf. on Acoustics, Speech and Signal Processing ICASSP85, Tampa, FL (1985) pp. 1037–1040.

TIME-DEPENDENT SPECTRA FOR
NON-STATIONARY STOCHASTIC PROCESSES

P. Flandrin
Laboratoire de Treitement du Signal, Lyon, France

Abstract

An overview of non-parametric approaches to time-dependent spectral analysis of non-stationary stochastic processes is provided. Possible definitions, which are based either on the spectral representation of the processes or on their covariance function, are derived from *a priori* requirements. Their respective merits, concerning both theoretical and practical properties, are evaluated and compared by means of typical examples. Estimation procedures are addressed and the usefulness of time-frequency descriptions for gaining new insights in some statistical Signal Processing problems is discussed.

1. Introduction

Among all the available tools in Signal Processing, spectral analysis is of special importance. This is due to the rather universal character of the major concept on which spectral analysis relies : that of *frequency*. If we consider indeed domains concerned with either physical waves (acoustics, vibrations, geophysics, optics, ...) or periodical events (econometrics, biology, astronomy, ...), a frequency description is generally basic for a better understanding of the phenomena which are involved. This has led to sustained efforts for developing efficient algorithms (and, hence, software packages and specialized processors), making of spectral analysis an usual and somewhat trivial way of (pre-)processing signals.

However, an even superficial inspection of "real life" signals imposes to fix limits and to raise objections to a systematic use of spectral analysis since it relies, through the concept of frequency itself, upon an implicit assumption of steady-state behavior, or *stationarity*. In fact, most of the natural signals are *non-stationary* and, strictly speaking, they do not enter the theoretical framework of spectral analysis. We are then faced with a contradiction between a large development of powerful techniques and their restricted (strict) applicability. This motivates the need for descriptions which take into account frequency aspects, which are nevertheless known to be physically relevant, together with a possible time-dependence, which is imposed by realistic considerations.

In order to derive such *time-frequency distributions* or *time-dependent spectra* , a first approach could be to consider local approximations for which the classical (stationary) notion of spectrum is valid. Owing to its clear limitations, this approach will not be followed up here. A less *ad hoc* approach will be preferred, which considers a non-stationary process as such, and tries to attach to it a well-defined function of both time and frequency. Furthermore, emphasis will be made on *non-parametric* methods in order to relax, as much as possible, the *a priori* assumptions which are always present in any modeling technique. At last, only *stochastic* processes will be considered.

The paper is organized as follows. Section 2 is devoted to the characteristic properties and concepts related to stationarity. This allows to precise in Section 3 what a time-dependent spectrum should be, in order to extend the notion of spectrum to the non-stationary case. Different solutions are then discussed and criticized, and Section 4 is concerned with a thorough discussion of the properties of the retained definition : the Wigner-Ville spectrum. Estimation procedures are tackled in Section 5, whereas Section 6 emphasizes the usefulness of time-frequency distributions for a natural and physically meaningful formulation of some statistical Signal Processing problems.

2. Stationarity

(All the material presented in this Section can be found in classical textbooks (e.g. : Papoulis, 1965; Wong, 1971; Koopmans, 1974; Blanc-Lapierre and Picinbono, 1981; Priestley, 1981), and we refer the reader to them for further details).

2.1. Weakly stationary processes

Since non-stationarity refers to a negative property, its simplest definition is related to that of the corresponding positive property. In a very general manner, a stochastic process is said to be *(strictly) stationary* if its probabilistic behavior remains the same throughout all time. This means that *all* its finite-dimensional probability distributions are time-invariant and, therefore, only depend on time differences. Strict stationarity is of course a very strong assumption and, in practical situations, it is very often more reasonable to only consider weak stationarity. By definition, a process x(t) is said to be *weakly* (or *wide-sense*, or *second-order*) *stationary* if

(2-1) $E[x(t)] = m_x = \text{constant}$;

(2-2) $E[(x(t_1) - m_x)(x(t_2) - m_x)^*] = r_x(t_1, t_2) = \gamma_x(t_1 - t_2)$;

(2-3) $\text{Var}[x(t)] = \gamma_x(0) < \infty$.

(where E stands for the expectation operator and the star for the complex conjugation), i.e. if stationarity (understood as time unchangingness) is only satisfied up to the second order.

Except in the Gaussian case, for which a probability distribution is entirely determined by its first and second moments, weakly stationary processes need not be strictly stationary. Nevertheless, and for a sake of brevity, *stationarity* will always stand for *weak stationarity* in the sequel of the paper and, unless otherwise mentioned, we will restrict, without loss of generality, to *zero-mean* processes.

2.2. Time-domain characterization

In discrete-time ($t = \ldots, -1, 0, +1, \ldots$), real-valued, zero-mean, *non-deterministic* stationary processes admit a decomposition, the so-called *Wold's decomposition*, according to which they can be represented as one-sided moving average (MA) processes, the order of which may be infinite. This expresses as

$$(2\text{-}4) \qquad x(t) = \sum_{s=-\infty}^{t} h(t-s)\,\varepsilon(s) \quad ; \quad E[\varepsilon(s_1)\,\varepsilon(s_2)] = \delta(s_1 - s_2) \quad ; \quad \sum_{s=0}^{\infty} h^2(s) < \infty \, ,$$

where $h(t)$ is the impulse response of a (time-invariant) MA filter and $\varepsilon(t)$ a discrete-time white noise of unit variance.

It follows directly from the above decomposition that the covariance function of a stationary process can be written as

$$(2\text{-}5) \qquad r_x(t_1, t_2) = \sum_{u=0}^{\infty} h(u)\,h(u + |t_1 - t_2|)$$

and, hence, only depends on time differences.

The continuous-case is more involved, but we shall admit that a continuous-time counterpart of (2-4) is, under reasonable conditions,

$$(2\text{-}6) \qquad x(t) = \int_{-\infty}^{t} h(t-s)\,dB(s) \quad ; \quad E[dB(s_1)\,dB(s_2)] = \delta(s_1 - s_2)\,ds_1\,ds_2 \, ,$$

where $B(t)$ is ordinary Brownian motion with unit strength.

2.3. Spectral representation

Beside this time-domain decomposition, stationary processes also admit a spectral representation of the form

$$(2\text{-}7) \qquad x(t) = \frac{1}{2\pi} \int_{-\infty}^{+\infty} e^{i\omega t}\,dX(\omega) \, ,$$

if the process is defined in continuous-time (in the discrete-time case, the integration interval should be $] - \pi, +\pi\,]$).

Such a decomposition possesses the remarkable advantage of being *doubly orthogonal* in the sense that

(2-8a) $$\int_{-\infty}^{+\infty} e^{i(\omega_1 - \omega_2)t} \, dt = 2\pi \, \delta(\omega_1 - \omega_2)$$

and

(2-8b) $E[dX(\omega_1) \, dX^*(\omega_2)] = 2\pi \, \delta(\omega_1 - \omega_2) \, dS_x(\omega_1) \, d\omega_2$.

This means that the basis functions of the decomposition are orthogonal w.r.t. the usual scalar product of functions on the real line, whereas the spectral increments are uncorrelated. Furthermore, the presence of complex exponentials as basis functions matches the mathematical simplicity of a doubly orthogonal decomposition with a nice physical interpretation in terms of (angular) *frequency*.

The decomposition (2-7) of the process itself carries over to that of its covariance function since we have from (2-8b)

(2-9) $$r_x(t_1, t_2) = \frac{1}{2\pi} \int_{-\infty}^{+\infty} e^{i\omega(t_1 - t_2)} \, dS_x(\omega) .$$

Again, we can check that stationary processes are characterized by a covariance function independent of absolute time. If we suppose then that the spectral measure is absolutely continuous w.r.t. Lebesgue measure, i.e. that it can be written as

(2-10) $dS_x(\omega) = \Gamma_x(\omega) \, d\omega$,

we obtain directly

(2-11) $r_x(t_1, t_2) = \gamma_x(t_1 - t_2)$,

where $\gamma_x(t)$ is the Fourier transform of $\Gamma_x(\omega)$.

Since, by construction, covariance functions are non-negative definite functions, it follows that $\Gamma_x(\omega)$ is everywhere non-negative : it is interpreted as the *power spectral density* of the process (Wiener-Khintchine's theorem).

Beside the physical interpretation of frequency, the spectral representation of stationary processes is of special importance in linear filtering operations. If we consider the process

$$(2\text{-}12) \qquad y(t) = \int_{-\infty}^{+\infty} h(s)\, x(t-s)\, ds \,,$$

resulting from the linear filtering (by a filter of impulse response $h(t)$ and transfer function $H(\omega)$) of a stationary process $x(t)$, we obtain the decomposition

$$(2\text{-}13) \qquad y(t) = \frac{1}{2\pi} \int_{-\infty}^{+\infty} e^{i\omega t}\, [H(\omega)\, dX(\omega)] \,.$$

It follows that the input-output relationship for linear filters and stationary processes takes on the simple form

$$(2\text{-}14) \qquad dS_y(\omega) = |\, H(\omega)\,|^2\, dS_x(\omega) \,.$$

This means that the spectral behavior of the output is that of the input, weighted by the squared modulus of the transfer function : this is a direct consequence of the fact that the complex exponentials are the eigenfunctions of linear time-invariant filters, considered as linear operators.

2.4. Non-stationary processes

By definition, *non-stationary* processes are those stochastic processes for which the characteristic stationarity properties considered previously are no more valid. As a first consequence, and in contrast with the stationary case, non-stationary processes do not admit a doubly orthogonal decomposition based on complex exponentials. We will discuss this in more detail in Section 3, but we can yet point out that departing from stationarity will necessarily lead to relax either the usual concept of frequency or the uncorrelation between spectral increments. Of course, considering second-order properties, the study of non-stationary processes will need to deal not only with time differences, but also with absolute time.

Among all the possible non-stationary processes, some are of special interest for their relations with stationary processes, and the intuitive interpretation which can be attached to them. As a first example, we can consider *locally stationary* processes (Silverman, 1957) defined, via their covariance function, by a factorization of the type

$$(2-15) \quad r_x(t_1, t_2) = m_x(\frac{1}{2}(t_1 + t_2)) \, \gamma_x(t_1 - t_2) ,$$

where $m_x(t)$ is a non-negative function and $\gamma_x(t)$ a non-negative definite function (stationary covariance function). If the function $m_x(t)$ should happen to be constant, it is clear that locally stationary processes would reduce to (weakly) stationary ones, whereas, in the general case, the formulation (2-15) takes into account a possible time evolution of a "local" stationarity : this is more apparent when (2-15) is rewritten under the symmetrized form

$$(2-16) \quad r_x(t + \frac{\tau}{2}, t - \frac{\tau}{2}) = m_x(t) \, \gamma_x(\tau) .$$

A second example is provided by *uniformly modulated* processes (Priestley, 1981) constructed as

$$(2-17) \quad x(t) = c(t) \, x_0(t) ,$$

where $c(t)$ is a (positive) modulation function, and $x_0(t)$ a stationary process. Such processes, which intuitively model time-dependent fluctuations in power, are clearly non-stationary since we have

$$(2-18) \quad r_x(t_1, t_2) = c(t_1) \, c(t_2) \, \gamma_{x_0}(t_1 - t_2) .$$

Physically, if the time evolution of the modulation function $c(t)$ is negligible when compared to the essential support of the covariance function of $x_0(t)$, we have on this support

$$(2-19) \quad c(t_1) \, c(t_2) \approx c^2(\frac{1}{2}(t_1 + t_2))$$

and such uniformly modulated processes are also locally stationary ones.

These two examples will be used, among other ones, throughout the paper, since they present the advantage of giving some insight into the transition from stationarity to non-stationarity.

3. Definitions of time-dependent spectra

3.1. Desirable properties

The frequency content of stationary processes is correctly depicted by the *power spectral density* which, as a function of frequency only, is well-suited for time-unchanging second-order properties. Nevertheless, as soon as non-stationary processes are concerned, new tools are to be defined for taking into account a possible time evolution of their frequency content. In order to achieve such definitions of *time-frequency distributions* or *time-dependent spectra* , a possible approach is first to impose some requirements which are to be fulfilled by the distribution that we are looking for, and second to find satisfactory candidates. This selective approach goes back to Blanc-Lapierre and Picinbono, 1955 and Loynes, 1968; it has been more recently revived by Martin, 1982b, Grenier, 1984-1987 and Flandrin, 1987a. This has led to different lists of required properties that we will not reproduce here, focusing rather on their general underlying philosophy : <u>time-dependent spectra are supposed to provide a natural extension, with an explicit time-dependence, of the classical notion of power spectral density, together with most of its nice properties</u>.

An implicit consequence of this is that the usual concept of frequency, stemming from the stationary case, should be preserved when passing from the power spectral density to a time-dependent spectrum.

Therefore, if $\rho_X(t, \omega)$ stands for the wanted time-dependent spectrum, a first class of reasonable constraints is the following :

1. *$\rho_X(t, \omega)$ is a real-valued function of time and frequency, everywhere non-negative.*

This should ensure the time-dependent spectrum to share with the power spectral density a physical interpretation of density along the frequency axis, while possessing a time-dependence.

Keeping in mind this aspect of time-dependent spectrum understood as "time unfolding" of the stationary power spectral density, a second class of constraints could be the following :

2. $\rho_X(t, \omega)$ *reduces to the ordinary power spectral density if the process is stationary*

(3-1) $r_x(t_1, t_2) = \gamma_x(t_1 - t_2) \Rightarrow \rho_x(t, \omega) = \Gamma_x(\omega)$,

and its marginal distribution in time (resp. frequency) reflects the second-order properties of the process in time (resp. frequency).

At last, the postulated time-dependent spectrum should satisfy some invariance properties related to usual transformations in Signal Processing, and again in direct extension of the stationary case. This leads to a third possible class of constraints, e.g. :

3. $\rho_X(t, \omega)$ *is compatible with time and frequency shifts*

(3-2) $y(t) = x(t - t_0) e^{i\omega_0 t} \Rightarrow \rho_y(t, \omega) = \rho_x(t - t_0, \omega - \omega_0)$;

it is support-preserving in both time and frequency, and it provides easy and meaningful input-output relationships for linear filters.

This third class is by no way exhaustive and it could be extended to cover e.g. scale invariance, scalar product conservation, time reversal compatibility, ...

Beside these requirements which can be thought of as *theoretical* ones and which can be handled in a rather *objective* way, there exist also more *practical* desirable properties which are more or less implicitly supposed to hold when looking for a time-dependent spectrum. These can read for instance :

4. $\rho_X(t, \omega)$ *provides a satisfactory picture of non-stationarities in the time-frequency plane, and it can be efficiently estimated from one observed realization of the process.*

Although such a constraint is of great importance for concrete implementations of time-frequency analyses, it is clear that its intuitive but somewhat fuzzy formulation calls for more *subjective* answers.

All the desired constraints being specified, the question is now to find suitable candidates for fulfilling them. At least two different possibilities are offered, depending whether the time-dependent spectrum is constructed from the covariance function or from the process itself.

In the first case, a possible and reasonable assumption is to impose a linear relation between $\rho_x(t, \omega)$ and $r_x(t_1, t_2)$ (Loynes, 1968). Such a connection can take on the form

$$(3-3) \qquad \rho_x(t, \omega) = \iint\limits_{-\infty}^{+\infty} D(t_1, t_2; t, \omega)\, r_x(t_1, t_2)\, dt_1\, dt_2 \,.$$

where D is some kernel function. A remarkable result is then that the shift invariance property (3-2) is sufficient to reduce admissible kernels to those of the form

$$(3-4) \qquad D(t_1, t_2; t, \omega) = F(\frac{1}{2}(t_1 + t_2) - t, t_1 - t_2)\, e^{-i\omega(t_1 - t_2)} \,,$$

where F is some arbitrary function. This leads to the stochastic counterpart of the general class of time-frequency distributions for deterministic signals, usually referred to as *Cohen's class* (Cohen, 1966).

A second example of constraint in the above linear model is the transition to stationarity. It follows directly from (3-3) that (3-1) is satisfied when

$$(3-5) \qquad \int\limits_{-\infty}^{+\infty} D(\theta + \frac{\tau}{2}, \theta - \frac{\tau}{2}; t, \omega)\, d\theta = e^{-i\omega\tau} \,,$$

which, in the shift-invariance case (3-4), reduces to

$$(3-6) \qquad f(0, \tau) = 1 \,,$$

where f is the Fourier transform of F on the first variable.

In fact, the situation which consists in imposing various constraints within Cohen's class, and translating them into properties of the arbitrary kernel function, is somewhat equivalent to that practiced in the deterministic case (Claasen and Mecklenbrauker, 1980c). It is then possible to import from the deterministic case most of its results and especially the most important one, according to which there exist some *incompatibilities between constraints*. A proof of this claim can be found in Wigner, 1971, keeping in mind that it has been long ago conjectured to hold (Blanc-

Lapierre and Picinbono, 1955; Loynes, 1968). In fact, we can maintain that there is no time-dependent spectrum, linearly connected to the covariance function, which satisfies all the required properties. The consequence of this is twofold :

1. *There is no chance to obtain one unique and well-defined time-dependent spectrum ;*

2. *For each chosen definition, it will be necessary to drop at least one desirable property.*

Furthermore, it can be shown that the most critical property is *positivity*, in the sense that imposing positivity is exclusive of almost all the other required properties (Janssen, 1988).

The second possibility of definition mentioned above was to construct a time-dependent spectrum directly from the process itself. If we go back to the doubly orthogonal decomposition (2-7), which is characteristic of stationary processes, it appears that a process-based definition of a time-dependent spectrum needs to modify (2-7) for extending it to non-stationary processes. Again, we are faced with two different approaches :

1. The first one is *to preserve the double orthogonality* of the decomposition, but this can only be achieved by relaxing the choice of complex exponentials as basis functions, and hence, to loose the concept of frequency ;

2. The second one is *to preserve the classical (stationary) concept of frequency* , but necessarily to accept some correlation between spectral increments.

These two approaches will now be discussed, but, as we can see, and in accordance with the covariance-based construction, there is no completely satisfactory solution to the addressed problem, justifying in some sense the known multiplicity of candidates which have been proposed in the literature since almost half a century.

3.2. Orthogonality preserving solutions

3.2.1. Karhunen's decompositions

In order to preserve a doubly orthogonal decomposition, a solution is to replace the complex exponentials by other basis functions $\Psi(t, \omega)$ such that

$$(3\text{-}7) \qquad x(t) = \frac{1}{2\pi} \int_{-\infty}^{+\infty} \Psi(t, \omega) \, dX(\omega) ,$$

with

$$(3\text{-}8) \qquad \int_{-\infty}^{+\infty} \Psi(t, \omega_1) \, \Psi^*(t, \omega_2) \, dt = 2\pi \, \delta(\omega_1 - \omega_2) .$$

Such solutions exist for second-order processes (Blanc-Lapierre and Picinbono, 1981) and the resulting (Karhunen's) decompositions are doubly orthogonal, but the variable ω *has no reason to be interpreted as frequency.*

From the decomposition of the process, we can derive the decomposition of the covariance function

$$(3\text{-}9) \qquad r_x(t_1, t_2) = \frac{1}{2\pi} \int_{-\infty}^{+\infty} \Psi(t_1, \omega) \, \Psi^*(t_2, \omega) \, dS_x(\omega) ,$$

from which it follows that

$$(3\text{-}10) \qquad \mathrm{Var}[x(t)] = \frac{1}{2\pi} \int_{-\infty}^{+\infty} | \Psi(t, \omega) |^2 \, dS_x(\omega) .$$

The interpretation of such a decomposition is then that each "spectral" contribution is weighted by a time-dependent function. This allows to formally define a time-dependent "spectrum" by

$$(3\text{-}11) \qquad dI_{x;t}(\omega) = | \Psi(t, \omega) |^2 \, dS_x(\omega) ,$$

which, in the case of absolute continuity w.r.t. Lebesgue measure :

$$(3\text{-}12) \qquad dI_{x;t}(\omega) = K_x(t, \omega) \, d\omega \quad ; \quad dS_x(\omega) = \Gamma_x(\omega) \, d\omega ,$$

reduces to

$$(3\text{-}13) \qquad K_x(t, \omega) = | \Psi(t, \omega) |^2 \, \Gamma_x(\omega) .$$

This is certainly a non-negative, real-valued and time-dependent function which reduces to the ordinary power spectral density in the case of stationary processes, but its spectral interpretation can be questioned.

Nevertheless, two very interesting relatives can be derived from it, which will be now discussed.

3.2.2. Priestley's evolutionary spectra

In his original work (Priestley, 1965), Priestley considered decompositions of the type (3-7) with

$$(3-14) \qquad \Psi(t, \omega) = A(t, \omega)\, e^{i\omega t} .$$

Processes which admit such (non-unique) representations are called *oscillatory* processes. Physically, they are aimed at describing temporal evolutions of spectral contributions, frequency by frequency. If the $A(t, \omega)$ are slowly varying functions in time, the introduction of oscillatory processes achieves some trade-off between orthogonality and frequency interpretation, since the basis functions $\Psi(t, \omega)$ are then almost orthogonal and, although the variable ω is not really defined as the usual frequency, its physical interpretation is very similar. In order to satisfy a "quasi-frequency" interpretation, the admissible modulation functions are then supposed to possess a representation

$$(3-15) \qquad A(t, \omega) = \frac{1}{2\pi} \int_{-\infty}^{+\infty} e^{itn}\, dK_\omega(n) .$$

with $|\, dK_\omega(n)\,|$ having an absolute maximum at $n = 0$.

We deduce from (3-11) and (3-14) that the corresponding time-dependent spectra (*Priestley's evolutionary spectra*) take on the form

$$(3-16) \qquad dH_{x;\,t}(\omega) = |\, A(t, \omega)\,|^2\, dS_x(\omega) ,$$

which leads to

$$(3-17) \qquad P_x(\omega) = |\, A(t, \omega)\,|^2\, \Gamma_x(\omega) ,$$

in the case (3-12) for which

(3-18) $dH_{x;\,t}(\omega) = P_x(t, \omega)\,d\omega$.

Priestley's evolutionary spectra certainly possess numerous advantages : they correspond to real-valued, non-negative functions, which restrict naturally to the ordinary power spectral density in the case of stationary processes, since we have then $A(t, \omega) = 1$.

Moreover, as far as the frequency interpretation of ω is valid, they provide us with a satisfactory instantaneous description of spectral properties. This can be illustrated by the example of uniformly modulated processes (2-17) for which we obtain

(3-19) $dH_{x;\,t}(\omega) = c^2(t)\,dS_{x_0}(\omega)$.

For any non-stationary process, the marginal distribution (in time) of Priestley's evolutionary spectra is meaningful since

(3-20) $\dfrac{1}{2\pi}\displaystyle\int_{-\infty}^{+\infty} dH_{x;\,t}(\omega) = Var[x(t)]$

and, furthermore, if some slow variation assumptions are satisfied, they furnish a natural (approximate) extension of input-output relationships for linear filters (Priestley, 1981).

Unfortunately, Priestley's evolutionary spectra do possess also some drawbacks which have limited their use. First, from a point of view of *definition*, one weakness, which has been already noted, is related to the approximate frequency interpretation. Another problem concerns the underlying class of oscillatory processes, which has been shown to be not so well-defined : for instance, it is very difficult for a given process to be declared as belonging or not to the class of oscillatory processes and, furthermore, this class is not closed under addition (Battaglia, 1979). Second, from a point of view of *estimation*, we will see in Section 5 that the proposed estimators come up with classical quasi-stationary procedures (related to complex demodulation or short-time periodograms), sacrifying hence most of the advantages of the theoretical and original definition.

The introduction of Priestley's evolutionary spectra was performed in the "frequency" domain. However, a time-domain interpretation is also possible, which gives some insight into its interpretation. Let us suppose that the modulation function possesses a Fourier decomposition

(3-21) $A(t, \omega) = \displaystyle\int_{-\infty}^{+\infty} h(t, s)\, e^{-i\omega s}\, ds$.

From the representation (2-7), we can construct a stationary process $x_0(t)$ by

(3-22) $x_0(t) = \dfrac{1}{2\pi} \displaystyle\int_{-\infty}^{+\infty} e^{i\omega t}\, dX(\omega)$.

It follows then from (3-14) and (3-21) that (3-7) can be written as

(3-23) $x(t) = \displaystyle\int_{-\infty}^{+\infty} h(t, s)\, x_0(t - s)\, ds$.

This means that a non-stationary oscillatory process can be viewed as the output of a linear time-variant filter (whose impulse response is the Fourier transform of the modulation function), when excited by a stationary input. This interpretation, which is physically very interesting, suggests a novel approach, starting directly from a time-domain decomposition.

3.2.3. The evolutive spectrum of Tjøstheim and Mélard

The general idea is now to start from *Cramér's decomposition* of non-stationary processes, extension of Wold's decomposition of stationary processes. A rigorous and canonical derivation of the resulting definition is due to Tjøstheim, 1976 and can be found e.g. in Grenier, 1987. For a sake of simplicity, we will restrict here to the more unformal presentation by Mélard, 1978, and to the discrete-time case.

In the non-stationary case, Wold's decomposition (2-4) can be extended to (Cramér, 1961)

(3-24) $x(t) = \displaystyle\sum_{s=-\infty}^{t} h(t, s)\, \varepsilon(s)$; $E[\varepsilon(s_1)\, \varepsilon(s_2)] = \delta(s_1 - s_2)$,

where $\varepsilon(t)$ is always supposed to be a stationary white noise of unit variance. From Cramér's decomposition, a non-stationary process is thus interpreted physically as the output of a linear time-variant filter excited by white noise.

Since $\varepsilon(t)$ is stationary, it follows from (2-7) that it admits a decomposition of the type

$$(3-25) \quad \varepsilon(s) = \frac{1}{2\pi} \int_{-\pi}^{+\pi} e^{i\omega s} \, dE(\omega) \; ; \; E[dE(\omega_1) \, dE^*(\omega_2)] = 2\pi \, \delta(\omega_1 - \omega_2) \, d\omega_1 \, d\omega_2 \, .$$

Inserting then (3-25) into (3-24), we simply obtain

$$(3-26) \quad x(t) = \frac{1}{2\pi} \int_{-\pi}^{+\pi} \Psi(t, \omega) \, dE(\omega) \, ,$$

where the basis functions are of the form

$$(3-27) \quad \Psi(t, \omega) = \sum_{s=-\infty}^{t} h(t, s) \, e^{i\omega s} \, .$$

The decomposition (3-26) is, in discrete-time, a special case of (3-7) and, hence, it gives access to a time-dependent spectrum, the *evolutive spectrum of Tjøstheim and Mélard*, defined by

$$(3-28) \quad TM_x(t, \omega) = | \sum_{s=-\infty}^{t} h(t, s) \, e^{i\omega s} |^2 \, .$$

This evolutive spectrum certainly possesses some advantages. It is a real-valued, non-negative quantity which reduces to the ordinary power spectral density when the process is stationary. Furthermore, its marginal distribution (in time) is meaningful since

$$(3-29) \quad \int_{-\pi}^{+\pi} TM_x(t, \omega) \, \frac{d\omega}{2\pi} = Var[x(t)] \, ,$$

and the time and frequency shift invariance property holds.

In the case of uniformly modulated processes (2-17), the (time-variant) impulse response of (3-24) admits a representation of the type

$$(3-30) \quad h(t, s) = c(t) \, h_0(t - s) \, ,$$

where $h_0(t)$ is the (time-invariant) impulse response associated to the stationary process $x_0(t)$.

Hence :

$$(3-31) \qquad TM_x(t, \omega) = c^2(t)\, \Gamma_{x_0}(\omega)$$

and, again, we obtain a satisfactory description of "instantaneous" spectral properties.

In contrast with this example, there are cases for which the obtained representation leads to unsatisfactory results, which are mostly due to the causality of the basically MA representation of the process. For instance, if we consider (Mélard, 1978) the piecewise stationary process

$$(3-32) \qquad x(t) = \begin{cases} x_1(t) \; ; \; t < 0 \\ x_2(t) \; ; \; t \geq 0 \end{cases}$$

where $x_1(t)$ and $x_2(t)$ are two different stationary processes of respective power spectral densities $\Gamma_1(\omega)$ and $\Gamma_2(\omega)$, we obtain

$$(3-33) \qquad TM_x(t, \omega) = \Gamma_1(\omega) \; ; \; t < 0 \, ,$$

but only

$$(3-34) \qquad \lim_{t \to \infty} TM_x(t, \omega) = \Gamma_2(\omega) \, .$$

Moreover, it can be shown that there is no one-to-one mapping between the covariance function and the evolutive spectrum (Mélard, 1978).

One way out for improving this situation is, when possible, to realize infinite order MA models of the type (3-24) in terms of finite-dimensional state-space models. This leads to the *rational evolutionary spectrum*, or *Grenier's relief*, which is based on time-dependent ARMA modeling and can be thought of as the rational spectrum of the stationary process which is, at each time point, tangential to the considered non-stationary process. We will not go further in this direction here, and refer the interested reader to Grenier, 1983-1984-1987.

3.3. Frequency preserving solutions

3.3.1. Loève's decomposition

In the non-stationary case, the doubly orthogonal decomposition (2-7), which was valid for the stationary case, can now be replaced in some cases by a similar expression

$$(3-35) \quad x(t) = \frac{1}{2\pi} \int_{-\infty}^{+\infty} e^{i\omega t} \, dX(\omega) \, ,$$

where the spectral increments are *no more uncorrelated*:

$$(3-36) \quad E[dX(\omega_1) \, dX^*(\omega_2)] = \Phi_x(\omega_1, \omega_2) \, d\omega_1 \, d\omega_2 \, .$$

This means that the corresponding non-stationary processes are characterized, in the frequency domain, by means of a *two-dimensional spectral distribution function* $\Phi_x(\omega_1, \omega_2)$ which is not perfectly concentrated along the diagonal $\omega_1 = \omega_2$. This contrasts with the stationary case for which

$$(3-37) \quad \Phi_x(\omega_1, \omega_2) = 2\pi \, \delta(\omega_1 - \omega_2) \, \Gamma_x(\omega_1) \, .$$

A decomposition like (3-35) has the drawback of being constructed on correlated increments, but it presents the considerable advantage of preserving the concept of (physical) frequency which stemms from the use of complex exponentials as basis functions.

3.3.2. Harmonizability

The condition under which the decomposition (3-35) exists is given by *Loève's condition* (Loève, 1962) :

$$(3-38) \quad \iint_{-\infty}^{+\infty} | \Phi_x(\omega_1, \omega_2) | \, d\omega_1 \, d\omega_2 < \infty \, .$$

Non-stationary processes which satisfy this condition are said to be *harmonizable*.

Harmonizability, which is a natural extension of weak stationarity, carries over to the covariance function which can then be written as

$$(3\text{-}39) \quad r_x(t_1, t_2) = \iint_{-\infty}^{+\infty} e^{i(\omega_1 t_1 - \omega_2 t_2)} \Phi_x(\omega_1, \omega_2) \frac{d\omega_1 \, d\omega_2}{4\pi^2}.$$

Therefore, the covariance function and the two-dimensional spectral distribution function form a two-dimensional Fourier transform pair which can be interpreted as the non-stationary counterpart of Wiener-Khintchine's relationship (2-9) between the stationary covariance function and the power spectral density, and which, of course, reduces to it in the stationary case.

3.3.3. The generalized Wigner-Ville spectrum

Since harmonizable processes admit two equivalent two-dimensional descriptions, one in the time domain and the other one in the frequency domain, it is natural to consider the wanted time-dependent spectrum as an intermediate description :

$$r_x(t_1, t_2) \quad \longleftrightarrow \quad \Phi_x(\omega_1, \omega_2)$$

$$\rho_x(t, \omega)$$

According to the Fourier connection between r_x and Φ_x, and a postulated linear relationship between r_x and ρ_x, the addressed problem reduces to that of finding a distribution which preserves the commutativity of the above diagram when the arrows represent Fourier transforms. This means that a relation of the type

$$(3\text{-}40) \quad r_x(t_1, t_2) = \int_{-\infty}^{+\infty} e^{i\omega\tau} \rho_x(t, \omega) \frac{d\omega}{2\pi}$$

(and a dual one for Φ_x) should hold for any linear transformation of the coordinates

$$(3\text{-}41) \quad \begin{pmatrix} t \\ \tau \end{pmatrix} = \begin{pmatrix} \alpha & \beta \\ \gamma & \delta \end{pmatrix} \begin{pmatrix} t_1 \\ t_2 \end{pmatrix}$$

which furthermore satisfies the isometry condition $| \alpha\delta - \beta\gamma | = 1$ for conserving energy under the transformations .

If we also impose the shift invariance property, it can be shown (Flandrin and Martin, 1983) that the resulting time-dependent spectrum takes on the general form (3-3)-(3-4) with the weighting function

$$(3\text{-}42) \qquad f_\sigma(n, \tau) = e^{i\sigma n\tau} \quad ; \quad \sigma = \alpha - \frac{1}{2} .$$

Labelling by Π_σ the two-dimensional Fourier transform of f_σ, we finally come up with two equivalent formulations of the admissible time-dependent spectra :

$$(3\text{-}43a) \qquad C_x(t, \omega; \Pi_\sigma) = \int_{-\infty}^{+\infty} r_x(t - (\sigma - \frac{1}{2})\tau, t - (\sigma + \frac{1}{2})\tau) \, e^{-i\omega\tau} \, d\tau ;$$

$$(3\text{-}43b) \qquad C_x(t, \omega; \Pi_\sigma) = \int_{-\infty}^{+\infty} \Phi_x(\omega - (\sigma + \frac{1}{2})n, \omega - (\sigma - \frac{1}{2})n) \, e^{-int} \frac{dn}{2\pi} .$$

This quantity is referred to as the *generalized Wigner-Ville spectrum*, in reference to the *Wigner-Ville distribution* of deterministic (non-random) signals

$$(3\text{-}44) \qquad W_x(t, \omega) = \int_{-\infty}^{+\infty} x(t + \frac{\tau}{2}) \, x^*(t - \frac{\tau}{2}) \, e^{-i\omega\tau} \, d\tau ,$$

introduced first in Quantum Mechanics by Wigner, 1932, rediscovered in Signal Theory by Ville, 1948, and recently popularized by the now classical papers of Claasen and Mecklenbrauker, 1980a-b-c.

The generalized Wigner-Ville spectrum only depends on one free parameter, and can then be specified by fixing it. Two particular cases are of special interest : if we impose $\sigma = 0$, we obtain the so-called *Wigner-Ville spectrum* (Martin, 1982a)

$$(3\text{-}45) \qquad W_x(t, \omega) = \int_{-\infty}^{+\infty} r_x(t + \frac{\tau}{2}, t - \frac{\tau}{2}) \, e^{-i\omega\tau} \, d\tau = \int_{-\infty}^{+\infty} \Phi_x(\omega - \frac{n}{2}, \omega + \frac{n}{2}) \, e^{-int} \frac{dn}{2\pi} ,$$

whereas the choice $\sigma = 1/2$ leads to *Rihaczek's spectrum*

$$(3\text{-}46) \quad R_x(t, \omega) = \int_{-\infty}^{+\infty} r_x(t, t - \tau)\, e^{-i\omega\tau}\, d\tau = \int_{-\infty}^{+\infty} \Phi_x(\omega - n, \omega)\, e^{-int}\, \frac{dn}{2\pi},$$

which is, for stochastic signals, the counterpart of *Rihaczek's distribution* (Rihaczek, 1968) of deterministic signals.

As defined in (3-43), the generalized Wigner-Ville spectrum possesses a considerable number of interesting properties (marginals, invariances, support property if $|\sigma| \leq 1/2$, ...), but also one drawback of great importance : in the general case, it takes on complex values, which certainly renders its physical interpretation difficult. Hence, we will first circumvent this difficulty before discussing in greater detail the properties of the admissible resulting time-dependent spectrum, which, in fact, will be shown to be the Wigner-Ville spectrum.

4. The Wigner-Ville spectrum

4.1. Conditional unicity

In the framework of the generalized Wigner-Ville spectrum, imposing the reality condition fixes directly the free parameter at the only value $\sigma = 0$, corresponding to the Wigner-Ville spectrum. This solution is then unique but, if we remember the impossibility for a time-dependent spectrum to satisfy simultaneously all the desirable properties, it is worthwhile to emphasize that such an unicity makes sense only for the corresponding specified set of constraints.

Apart from the approach developed here, different such sets can be found in Flandrin and Martin, 1983, which all lead to the Wigner-Ville spectrum as the unique admissible time-dependent spectrum. Nevertheless, special requirements can also be found, which result in other spectra : this is for instance the case if we impose a causality condition in the definition. In fact, if we assume that, at each time instant, the time-dependent spectrum only depends on past values of the process, it can be shown (Flandrin, Escudié and Martin, 1985) that the resulting solution takes on the form

$$(4\text{-}1) \quad 2\, \mathrm{Re}\, \{ \int_0^{+\infty} r_x(t, t - \tau)\, e^{-i\omega\tau}\, d\tau \},$$

referred to as *Page's spectrum* (Page, 1952; Grace, 1981).

As a second example, it has already been mentioned that the Wigner-Ville spectrum can very well take on negative values and, of course, imposing the positivity condition would certainly discard the Wigner-Ville spectrum as a suitable candidate. This suggests that the unicity of a time-dependent spectrum is, in some sense, matter of convenience; nevertheless, it will be justified in Section 4.3 sq that the Wigner-Ville spectrum really provides us with a satisfactory time-dependent spectrum when considering the largest number of desirable properties. Moreover, and apart from the theoretical constraints imposed until now, other important, but more subjective, requirements are to be taken into account for the choice of a proper time-dependent spectrum : it is for instance implicit but necessary to obtain *readable pictures* which clearly exhibit the non-stationarities of the analyzed process. In this respect also, the competitors to the Wigner-Ville spectrum can be shown not to be satisfactory (Flandrin, 1987a).

4.2. Relations with other time-dependent spectra

4.2.1. General covariance-based formulation

If we come back to the general form of shift-invariant covariance-based time-dependent spectra :

$$(4-2) \quad C_x(t, \omega; \Pi) = \iiint_{-\infty}^{+\infty} e^{in(u-t)} f(n, \tau) \, r_x(u + \frac{\tau}{2}, u - \frac{\tau}{2}) \, e^{-i\omega\tau} \frac{dn}{2\pi} \, du \, d\tau$$

(where f and Π are a two-dimensional Fourier transform pair), it is clear that different solutions will be provided when specifying the arbitrary function.

A first example was the generalized Wigner-Ville spectrum (3-42) and, as special cases, the Wigner-Ville spectrum itself or Rihaczek's spectrum.

Other time-dependent spectra can also be found as special cases, for instance Page's spectrum (4-1) characterized by

$$(4-3) \quad f(n, \tau) = e^{i(n|\tau|)/2}.$$

Any member of Cohen's class (4-2) is in fact linearly connected with the Wigner-Ville spectrum and we can write

$$(4-4) \qquad C_x(t, \omega; \Pi) = \iint\limits_{-\infty}^{+\infty} \Pi(t - u, \omega - n) \, W_x(u, n) \, du \, \frac{dn}{2\pi} .$$

This means that any covariance-based time-dependent spectrum can be thought of as a doubly convolved (in both time and frequency) Wigner-Ville spectrum. However, it must be noticed that this convolution is not necessarily of the form of a smoothing, or low-pass filtering operation (cf. e.g. the Rihaczek or the Page case).

4.2.2. Stochastic Wigner-Ville distribution

Given a stochastic process $x(t)$, we can imagine to construct a stochastic time-frequency distribution by direct application of the definition which holds in the deterministic case. This leads to the *stochastic Wigner-Ville distribution* defined by

$$(4-5) \qquad W_x(t, \omega) = \int\limits_{-\infty}^{+\infty} x(t + \frac{\tau}{2}) \, x^*(t - \frac{\tau}{2}) \, e^{-i\omega\tau} \, d\tau .$$

If we consider the quantity

$$(4-6) \qquad y_t(\tau) = x(t + \frac{\tau}{2}) \, x^*(t - \frac{\tau}{2}) ,$$

it can be shown (Martin, 1982a) that the condition

$$(4-7) \qquad | \iint\limits_{-\infty}^{+\infty} E[y_t(t_1) \, y_t^*(t_2)] \, e^{i(\omega_1 t_1 - \omega_2 t_2)} \, dt_1 \, dt_2 | < \infty$$

ensures that

$$(4-8) \qquad E[W_x(t, \omega)] = W_x(t, \omega) .$$

The Wigner-Ville *spectrum* appears then as the expectation value of the stochastic Wigner-Ville *distribution*.

4.2.3. Karhunen-based spectra

Connections can be established between the Wigner-Ville spectrum and Karhunen-based spectra. Starting from the representation (3-9) of the covariance function, we obtain

$$(4\text{-}9) \qquad W_x(t, \omega) = \iint_{-\infty}^{+\infty} \Psi(t + \frac{\tau}{2}, n) \, \Psi^*(t - \frac{\tau}{2}, n) \, e^{-i\omega\tau} \, \Gamma_x(n) \, d\tau \, \frac{dn}{2\pi}$$

and, hence, we have

$$(4\text{-}10a) \qquad W_x(t, \omega) = \int_{-\infty}^{+\infty} V(n; t, \omega) \, K_x(t, n) \, \frac{dn}{2\pi},$$

with

$$(4\text{-}10b) \qquad V_x(n; t, \omega) = |\Psi(t, n)|^{-2} \int_{-\infty}^{+\infty} \Psi(t + \frac{\tau}{2}, n) \, \Psi^*(t - \frac{\tau}{2}, n) \, e^{-i\omega\tau} \, d\tau.$$

A special case is that of Priestley's evolutionary spectra for which the relations (4-10) read (Hammond and Harrison, 1985)

$$(4\text{-}11a) \qquad W_x(t, \omega) = \int_{-\infty}^{+\infty} V(n; t, \omega) \, P_x(t, n) \, \frac{dn}{2\pi},$$

with

$$(4\text{-}11b) \qquad V_x(n; t, \omega) = |A(t, n)|^{-2} \int_{-\infty}^{+\infty} A(t + \frac{\tau}{2}, n) \, A^*(t - \frac{\tau}{2}, n) \, e^{-i(\omega - n)\tau} \, d\tau.$$

It follows from this expression that, as long as slowly varying processes are concerned, we may assume that

$$(4\text{-}12) \qquad A(t + \frac{\tau}{2}, n) \, A^*(t - \frac{\tau}{2}, n) \approx |A(t, n)|^2$$

for lags τ less than the equivalent width of the stationary covariance function $\gamma_x(\tau)$. This results in negligible differences between both spectra :

$$(4\text{-}13) \qquad W_x(t, \omega) \approx P_x(t, \omega).$$

4.2.4. Extension to non-harmonizable processes

Although the formulation (3-45) of the Wigner-Ville spectrum comes out from harmonizability, it can also be used as a *definition* for the time-dependent spectrum of *any* non-stationary process. This has been done long ago (Silverman, 1957, Mark, 1970, Bendat and Piersol, 1971, Bastiaans, 1981) and, furthermore, if we accept to deal with generalized functions, we can also relax the assumption of second-order processes. The resulting quantity certainly lacks from the suitable theoretical basis which is attached to harmonizable processes, but it allows to take into account useful idealizations such as *white noise*. For instance, if we consider a process $\varepsilon(t)$ with covariance function

(4-14) $r_\varepsilon(t_1, t_2) = \delta(t_1 - t_2)$,

we readily obtain

(4-15) $W_\varepsilon(t, \omega) = 1$,

which is exactly the wanted behavior for a time-dependent spectrum of white noise.

4.3. Properties of the Wigner-Ville spectrum

In this Section, we will enumerate most of the properties which are verified by the Wigner-Ville spectrum and make it attractive, in accordance with the requirements discussed in Section 3.1.

1. *The Wigner-Ville spectrum is a real-valued function of time and (physical) frequency.*

2. *The Wigner-Ville spectrum reduces to the classical power spectral density for stationary processes, and its marginal distributions in time and frequency reflect the second-order properties of the analyzed process, since they identify to its variance and its two-dimensional spectral distribution function along the principal diagonal :*

(4-16a) $\displaystyle \int_{-\infty}^{+\infty} W_x(t, \omega) \frac{d\omega}{2\pi} = \text{Var}[x(t)]$;

(4-16b) $\displaystyle\int_{-\infty}^{+\infty} W_x(t, \omega)\, dt = \Phi_x(\omega, \omega)$.

3. *By construction, the Wigner-Ville spectrum is shift-invariant. It is also support-preserving* :

(4-17a) $(x(t) = 0 \ , \ |t| > T) \Rightarrow (W_x(t, \omega) = 0 \ , \ |t| > T)$;

(4-17b) $(X(\omega) = 0 \ , \ |\omega| > \Omega) \Rightarrow (W_x(t, \omega) = 0 \ , \ |\omega| > \Omega)$;

and it provides a simple input-output relationship for linear filters, since the time-convolution structure is preserved, but as a function of frequency

(4-18a) $y(t) = \displaystyle\int_{-\infty}^{+\infty} h(s)\, x(t - s)\, ds \Rightarrow W_y(t, \omega) = \int_{-\infty}^{+\infty} W_h(s, \omega)\, W_x(t - s, \omega)\, ds$.

A dual relation holds for amplitude-modulated processes since

(4-18b) $y(t) = c(t)\, x(t) \Rightarrow W_y(t, \omega) = \displaystyle\int_{-\infty}^{+\infty} W_c(t, n)\, W_x(t, \omega - n)\, \dfrac{dn}{2\pi}$.

Concerning the point 4., it will be considered in some more detail in Section 5.2.

All these properties carry over to the discrete-time version of the Wigner-Ville spectrum

(4-19) $W_x(t, \omega) = 2 \displaystyle\sum_{k=-\infty}^{+\infty} r_x(t + k, t - k)\, e^{-i2\omega k}$,

where $t = ..., -1, 0, +1, ...$ and $|\omega| < \pi$. It has just to be noted that, as in the deterministic case, aliasing is avoided when using (4-19) if the analyzed process is either real-valued, but oversampled by a factor of at least two, or analytic (Claasen and Mecklenbrauker, 1980b).

4.4. Non-stationary characteristics derived from the definition

Apart from the properties listed above, the Wigner-Ville spectrum can also be used for *defining* useful quantities related to non-stationary characteristics.

A first example is given by the notion of *instantaneous frequency* $\omega_x(t)$ and *group delay* $t_x(\omega)$, which, in analogy to the deterministic case, can be defined as first-order local moments (or centroids) of the Wigner-Ville spectrum (Martin, 1982a) :

$$(4\text{-}20a) \quad \omega_x(t) = \int_{-\infty}^{+\infty} \omega\, W_{z_x}(t, \omega)\, \frac{d\omega}{2\pi} \Big/ \int_{-\infty}^{+\infty} W_{z_x}(t, \omega)\, \frac{d\omega}{2\pi},$$

where $z_x(t)$ is the analytic process associated to the real-valued one $x(t)$;

$$(4\text{-}20b) \quad t_x(\omega) = \int_{-\infty}^{+\infty} t\, W_x(t, \omega)\, dt \Big/ \int_{-\infty}^{+\infty} W_x(t, \omega)\, dt .$$

A second example is provided by a possible definition of an *average spectrum* on a time interval of length T (Flandrin, 1986c-1987a)

$$(4\text{-}21) \quad S_x(\omega; T) = \frac{1}{T} \int_0^T W_x(t, \omega)\, dt .$$

This takes a special importance for describing the global spectral behavior of processes with finite power by setting

$$(4\text{-}22) \quad S_x(\omega) = \lim_{T \to \infty} S_x(\omega; T) .$$

We obtain then a quantity very similar to the *Wiener spectrum* stemming from Generalized Harmonic Analysis (Wiener, 1930).

At last, the Wigner-Ville spectrum can be used for measuring the *degree of non-stationarity* of a stochastic process on a given time interval of length T. This can be achieved for instance by choosing as distance measure the quadratic quantity (Flandrin, 1987a)

$$(4\text{-}23) \quad d^2(\omega; T) = \frac{1}{T} \int_0^T W_x^2(t, \omega)\, dt - S_x^2(\omega; T) .$$

If we remember that the Wigner-Ville spectrum reduces to the ordinary power spectral density in the case of stationary processes, we see that stationarity is then characterized by a null distance, the scale being open. Of course, the distance need not be quadratic : an approach based on absolute deviations can be found in (Martin, 1984a).

4.5. Theoretical examples

4.5.1. Locally stationary and uniformly modulated processes

If we consider *locally stationary* processes characterized by (2-15), we obtain directly for the Wigner-Ville spectrum

$$(4-24) \quad W_x(t, \omega) = m_x(t) \, \Gamma_x(\omega) \, .$$

This expression depicts correctly a (local) time evolution of a stationary behavior. In the case of a time-dependent white noise

$$(4-25) \quad x(t) = c(t) \, \varepsilon(t) \quad ; \quad c(t) > 0 \quad ; \quad E[\varepsilon(t_1) \, \varepsilon(t_2)] = \delta(t_1 - t_2) \, ,$$

it reduces to the satisfactory expression (generalization of (4-15)) :

$$(4-26) \quad W_x(t, \omega) = c^2(t) \, .$$

This is to be compared to the more general class of *uniformly modulated* processes (2-17). In fact, if we make use of (4-18b), we obtain

$$(4-27) \quad W_x(t, \omega) = \int_{-\infty}^{+\infty} W_c(t, n) \, \Gamma_{x_0}(\omega - n) \, \frac{dn}{2\pi} \, .$$

Therefore, if the modulation function c(t) is slowly varying, i.e. if we assume that

$$(4-28) \quad c(t + \frac{\tau}{2}) \, c(t - \frac{\tau}{2}) \approx c^2(t) \quad , \quad |\tau| < \Theta_\gamma$$

(where Θ_γ is the equivalent width of the stationary covariance function of $x_0(t)$), we get the approximate form

$$(4-29) \quad W_x(t, \omega) \approx c^2(t) \, \Gamma_{x_0}(\omega) \, .$$

which is hence identical to the corresponding Priestley's evolutionary spectrum (3-19).

A companion example can be provided in discrete-time by the multiplicative MA(1) model :

$$(4\text{-}30) \quad x(t) = c(t) \, [\varepsilon(t) + b \, \varepsilon(t - 2)] \quad ; \quad E[\varepsilon(t_1) \, \varepsilon(t_2)] = \delta(t_1 - t_2) \, .$$

A straightforward calculation yields (Martin and Flandrin, 1985b)

$$(4\text{-}31) \quad W_x(t, \omega) = c^2(t) \, [(1 + b^2) + 2b \, c(t + 1) \, c^{-2}(t) \, c(t - 1) \cos 2\omega] \, ,$$

which is now to be compared with the corresponding evolutive spectrum of Tjøstheim and Melard :

$$(4\text{-}32) \quad TM_x(t, \omega) = c^2(t) \, [(1 + b^2) + 2b \cos 2\omega] \, .$$

Again, we can check that both spectra are similar if the slow variation condition

$$(4\text{-}33) \quad c(t + 1) \, c(t - 1) \approx c^2(t)$$

holds.

4.5.2. White noise driven systems

For non-stationary processes with multiplicity one (Cramer, 1971), the continuous-time counterpart of Cramer's decomposition (3-24) reads

$$(4\text{-}34) \quad x(t) = \int_{-\infty}^{t} h(t, s) \, dB(s) \quad ; \quad E[dB(s_1) \, dB(s_2)] = \delta(s_1 - s_2) \, ds_1 \, ds_2 \, ,$$

which extends the stationary case (2-6). We obtain hence for the Wigner-Ville spectrum

$$(4\text{-}35) \quad W_x(t, \omega) = \int_{-\infty}^{+\infty} [\int_{-\infty}^{t - |\tau|/2} h(t + \frac{\tau}{2}, s) \, h(t - \frac{\tau}{2}, s) \, ds] \, e^{-i\omega\tau} \, d\tau \, .$$

Such a result can be used for describing *transient* non-stationary processes like the response of linear filters to the onset of stationary white noise :

(4-36) $x(t) = \int_0^t h(t - s) \, dB(s) \cdot u(t)$,

where u(t) is the unit step function. In this case, (4-35) reduces to

(4-37) $W_x(t, \omega) = \int_0^t W_h(s, \omega) \, ds \cdot u(t)$.

If the impulse response is of finite duration T_h, we have thus

(4-38) $t > T_h \implies W_x(t, \omega) = | H(\omega) |^2$.

This means, in accordance with the support-preserving property, that the Wigner-Ville spectrum reduces to the power spectral density as soon as the transient (whose duration is that of the impulse response of the filter) is over.

4.5.3. Fractional Brownian motions

A special case related to the class (4-36) is that of *fractional Brownian motions* (Mandelbrot and van Ness, 1968) defined as

(4-39) $B_H(t) = \dfrac{1}{\Gamma(H+1/2)} [\int_{-\infty}^0 \{ |t - s|^{H-1/2} - | s |^{H-1/2} \} dB(s) + \int_0^t |t - s|^{H-1/2} dB(s)]$,

with 0 < H < 1. For such generalizations of ordinary Brownian motion (obtained for H = 1/2), it can then be shown (Flandrin, 1987c) that their Wigner-Ville spectrum expresses as

(4-40) $W_{B_H}(t, \omega) = (1 - 2^{1-2H} \cos \omega t) \cdot | \omega |^{-(1+2H)}$.

For this class of non-stationary processes, some properties of the Wigner-Ville spectrum, and concepts related to them, are of special interest. For instance, the concept of average spectrum finds here a very interesting application since we get

$$(4-41) \quad S_{B_H}(\omega) = |\omega|^{-(1+2H)}.$$

The use of the Wigner-Ville spectrum provides then us with a well-defined description of the global spectral behavior of these fractional Brownian motions, which are known empirically to obey power laws of fractional order although, due to their nonstationary structure, they do not possess any spectrum in the usual sense.

4.5.4. Shot noise

As a final example, we can consider *shot noise* s(t) defined by

$$(4-42) \quad s(t) = \sum_{k=-\infty}^{+\infty} h(t - t_k),$$

where the time instants t_k are Poisson distributed with a time-dependent density $\lambda(t)$ (Papoulis, 1965). We obtain then

$$(4-43) \quad W_s(t, \omega) = \int_{-\infty}^{+\infty} W_h(s, \omega) \lambda(t - s)\, ds$$

and, if we assume a slow variation of the density within the support of h(t), we have

$$(4-44) \quad W_s(t, \omega) \approx \lambda(t) \, | \, H(\omega) \, |^2 .$$

If we now suppose that

$$(4-45) \quad \lambda(t) = \lambda_0 \, u(t),$$

i.e. that a constant density is switched on at the time origin, we come up with

$$(4-46) \quad W_s(t, \omega) = \lambda_0 \int_0^t W_h(s, \omega)\, ds \cdot u(t).$$

The resulting Wigner-Ville spectrum is then identical to that of a white noise driven system, when the power spectral density of the white noise is replaced by the density of the Poisson impulses.

4.6. The positivity problem

4.6.1. Positivity condition and examples

After having considered all the properties of the Wigner-Ville spectrum and illustrated them on various theoretical examples, it turns out that almost only one desirable property is missing : *positivity*. Indeed, the Wigner-Ville spectrum has no reason to be everywhere non-negative, forbidding hence any density interpretation such as the one attached to the stationary concept of power spectral density.

However, if we keep in mind some of the examples discussed in the preceding Section, it appears that the possible occurence of negative values in the Wigner-Ville spectrum is not necessarily of dramatic importance. For instance, it is clear that any locally stationary process admits a non-negative Wigner-Ville spectrum since the right-hand side of (4-24) is everywhere non-negative.

More generally, a process is ensured to possess a non-negative Wigner-Ville spectrum if this latter is the Fourier transform of some non-negative definite function. Considering the decomposition (4-35), this means that the positivity condition reads (Flandrin, 1986b)

$$(4-47) \quad W_x(t, \omega) \geq 0 \quad \Leftrightarrow \quad \exists \, g(t, s) \quad /$$

$$\int_{-\infty}^{t - |\tau|/2} h(t + \frac{\tau}{2}, s) \, h(t - \frac{\tau}{2}, s) \, ds = \int_{-\infty}^{+\infty} g(t, s + \frac{\tau}{2}) \, g(t, s - \frac{\tau}{2}) \, ds \, .$$

This is for instance the case for ordinary Brownian motion characterized by

$$(4-48) \quad g(t, s) = \frac{1}{2} \sqrt{2} \, u(t) \, u(t - |s|) \, ,$$

and whose Wigner-Ville spectrum expresses as

$$(4-49) \quad W_{B_{1/2}} (t, \omega) = 2\, t^2 \left(\frac{\sin |\omega|t}{|\omega|t} \right)^2 u(t).$$

This holds also for all fractional Brownian motions with $1/2 \le H < 1$ (Flandrin, 1987c).

Other peculiar examples of processes with a non-negative Wigner-Ville spectrum can be provided in discrete-time (Flandrin, 1986b-c). A first one is the the multiplicative MA(1) model (4-30) for which a sufficient condition of positivity is

$$(4-50) \quad 0 < c \le c(t) \le C \quad ; \quad \left(\frac{C}{c} \right)^2 \le \left(\frac{1 + b^2}{2b} \right).$$

A second example is the time-dependent MA(1) model

$$(4-51) \quad x(t) = \varepsilon(t) + b(t - 2)\, \varepsilon(t - 2) \quad ; \quad E[\varepsilon(t_1)\varepsilon(t_2)] = \delta(t_1 - t_2)$$

with

$$(4-52a) \quad b(t) = b_0 + \delta \cos \omega_0 t$$

and the conditions

$$(4-52b) \quad 0 < b_0 < 1 \quad ; \quad \delta < \inf[b_0, (1 - \sqrt{b_0})^2].$$

In these two examples, positivity is certainly related to the processes themselves, which are constrained to lie not too far from stationarity. Nevertheless, these examples, along with other ones (Flandrin, 1986c), suggest that, in the stochastic case, positivity is by no way such an exception as in the deterministic case (Hudson, 1974). This can be interpreted as due also to the definition itself of the Wigner-Ville spectrum.

4.6.2. Probabilistic interpretation

In order to support the claim that positivity of the Wigner-Ville spectrum is intimately related to the underlying ensemble averaging, we can consider the following example (Janssen, 1988): a stochastic signal x(t) is obtained by affecting a deterministic (non-random) signal $x_d(t)$ of *jitter*, in both time and frequency, with a Gaussian probability density function

(4-53) $G(t, \omega) = e^{-(1/2)[(t/\Delta t)^2 + (\omega/\Delta \omega)^2]}$.

We have then

(4-54) $W_x(t, \omega) = \iint\limits_{-\infty}^{+\infty} G(t - \theta, \omega - n)\, W_{x_d}(\theta, n)\, d\theta\, \dfrac{dn}{2\pi}$

and it follows from a classical result (de Bruijn, 1967) that

(4-55) $\Delta t \cdot \Delta \omega \geq 1/2 \;\Rightarrow\; W_x(t, \omega) \geq 0$.

This means that a sufficient amount of randomness may ensure positivity.

A second example is that of a non-random signal $x_d(t)$ of finite energy E_d, embedded in stationary noise $b(t)$ of power spectral density $\Gamma_b(\omega)$:

(4-56) $x(t) = x_d(t) + b(t)$

for which

(4-57) $W_x(t, \omega) = W_{x_d}(t, \omega) + \Gamma_b(\omega)$.

Since we have always

(4-58) $|W_{x_d}(t, \omega)| \leq 2 E_d$,

it is clear that a sufficient condition of positivity reads

(4-59) $\Gamma_b(\omega) / 2 E_d \geq 1$.

The left-hand side of (4-59) being a measure of *noise-to-signal ratio*, this condition admits again the preceding interpretation.

5. Estimation of time-dependent spectra

The question which naturally follows the definition of time-dependent spectra is that of their estimation. In the common case where only one realization of a non-stationary process is available, the major problem is that ergodicity-based techniques, as those used in the stationary case, are generally no more valid. Some assumptions will then be necessary for any estimation procedure.

Focussing on non-parametric approaches, we will summarize them and the corresponding results in the case of both Priestley's evolutionary spectra and the Wigner-Ville spectrum. Furthermore, we will restrict ourselves to discrete-time processes observed on a finite duration , i.e. for time instants $t = 0, 1, ..., T$.

5.1. Priestley's evolutionary spectra and complex demodulation

Priestley has proposed (see e.g. Priestley, 1981) to only consider *semi-stationary* processes, i.e. processes for which there exists a finite quantity, called *characteristic width* and defined as

$$(5-1) \quad T_x = \sup \left\{ \left[\sup_{\omega} \left\{ \int_{-\infty}^{+\infty} |n| \, |dK_\omega(n)| \right\} \right]^{-1} \right\}.$$

The estimation is then achieved by means of a technique, called *complex demodulation* , which computes the quantity

$$(5-2) \quad S_x(t, \omega) = \sum_{u = -\infty}^{+\infty} h(u) \, x(t - u) \, e^{-i\omega(t - u)},$$

where $x(t)$ is the realization of the process and $h(t)$ is a normalized time window of equivalent width

$$(5-3) \quad T_h = \sum_{u = -\infty}^{+\infty} |u| \, |h(u)|$$

with the condition

$$(5-4) \quad T_h \ll T_x \ll T.$$

The principle of the procedure is to modulate the observed realization by a complex exponential, to pass the result through a low-pass filter (whose impulse response is h(t)), and to perform at last an envelope detection.

In fact, Priestley has shown that, within the above hypotheses, we have

$$(5\text{-}5) \quad E[\,|S_x(t, \omega)|^2\,] \approx \int_{-\pi}^{+\pi} |H(n)|^2\, P_x(t, \omega + n)\, \frac{dn}{2\pi}\,,$$

where $H(\omega)$ is the Fourier transform of $h(t)$. This shows that the proposed procedure provides us with a frequency smoothed version of the desired spectrum, whereas, due to the definition (5-2), the time resolution is of the order of T_h.

Such a procedure is then faced with a kind of *uncertainty principle* (Tsao, 1984), since any increase of resolution in the time (resp. frequency) direction can only be achieved at the expense of a corresponding decrease in the frequency (resp. time) direction.

Moreover, the variability of the estimate is very high, since, in the Gaussian case, it turns out that

$$(5\text{-}6) \quad \text{Var}[\,|S_x(t, \omega)|^2\,] \approx [\int_{-\pi}^{+\pi} |H(n)|^2\, P_x(t, \omega + n)\, \frac{dn}{2\pi}]^2\,.$$

This situation can be improved by introducing a time smoothing, according to the modified estimator

$$(5\text{-}7) \quad S'_x(t, \omega) = \sum_{s=-\infty}^{+\infty} g(s)\, |S_x(t + s, \omega)|^2\,.$$

This achieves variance reduction but, as a by-product, increases the bias in the time direction. Priestley's estimation procedure is then faced also with a necessary trade-off between bias and variance.

5.2. Estimation of the Wigner-Ville spectrum

5.2.1. General framework

If we consider the Wigner-Ville spectrum and its definition (4-19), the problem reduces to that of estimating the symmetrized covariance function. A possible approach is to restrict to processes undergoing a *slow evolution* in the sense that their covariance function admits, at each time instant and on a finite interval Θ, a local approximation by means of a (tangential) stationary covariance function $\gamma_{x;\,t}$. Assuming then that

(5-8) $| r_x(t + \tau, t - \tau) - \gamma_{x;\,t}(2\tau) | < \varepsilon(\Theta)$,

where $\varepsilon(\Theta)$ is a measure of the approximation, the largest value Θ_x for which (5-8) holds is referred to as the *time of stationarity* of the process (Martin and Flandrin, 1983-1985b).

Hence, a possible estimator of the time-dependent covariance function is

(5-9) $r_x^e(t + \tau, t - \tau) = \displaystyle\sum_{s=-\infty}^{+\infty} F(s, 2\tau)\, x(t + s + \tau)\, x^*(t + s - \tau)$,

where $x(t)$ is again the observed realization of the process and $F(s, 2\tau)$ is an arbitrary function, whose duration in s is supposed to be much smaller than the time of stationarity.

Inserting (5-9) into (4-19), it turns out that a general class of estimators for the Wigner-Ville spectrum reads (Flandrin and Martin, 1984)

(5-10) $W_x^e(t, \omega) = 2 \displaystyle\sum_{\tau=-\infty}^{+\infty} r_x^e(t + \tau, t - \tau)\, e^{-i2\omega\tau} = C_x(t, \omega; \Pi)$,

which is just the discrete-time form of the general time-frequency distribution (4-4) of the observed realization.

5.2.2. Statistical properties

Starting from the definition (5-10), general statistical properties of estimators of the Wigner-Ville spectrum can be derived, as parameterized by the arbitrary function Π. First and second order results are the following (Flandrin and Martin, 1984) :

$$(5-11) \quad E[W_x^e(t, \omega)] = \sum_{s=-\infty}^{+\infty} \int_{-\pi/2}^{+\pi/2} W_x(s, n) \, \Pi(t - s, \omega - n) \, \frac{dn}{2\pi}$$

and

$$(5-12) \quad \text{Cov}[W_x^e(t_1, \omega_1), W_x^e(t_2, \omega_2)] \approx$$

$$\sum_{s=-\infty}^{+\infty} \int_{-\pi/2}^{+\pi/2} W_x^2\left(\frac{t_1 + t_2}{2}, n\right) \Pi(s - t_1, \omega_1 - n) \, \Pi^*(s - t_2, \omega_2 - n) \, \frac{dn}{2\pi},$$

if we suppose that the local correlation time of the tangential stationary process is much smaller than the time of stationarity, and if the analyzed process is supposed to be Gaussian and analytic.

We deduce from (5-12) an expression for the approximate variance of the estimator, which reads

$$(5-13) \quad \text{Var}[W_x^e(t, \omega)] \approx \left[\sum_{s=-\infty}^{+\infty} \int_{-\pi/2}^{+\pi/2} |\Pi(s, n)|^2 \frac{dn}{2\pi} \right] W_x^2(t, \omega).$$

In the general case, we see that (5-10) provides us with estimators which are doubly biased and correlated, in both time and frequency. Nevertheless, since the corresponding performance is directly dependent on the function Π, it is now possible to handle special cases by fixing this function and, in return, to design satisfactory estimators by selecting suitable candidates.

5.2.3. Special cases and design of estimators

A first example of estimators are the so-called *short-time periodograms* (Allen and Rabiner, 1977) which have been already encountered in the estimation of Priestley's evolutionary spectra. In fact, it can be easily shown that the squared modulus of (5-2) is a special case of (5-10), when choosing the arbitrary function Π as the Wigner-Ville distribution of the observation window $h(t)$. We have then directly

$$(5-14) \quad E[\, |S_x(t, \omega)|^2 \,] = \sum_{s=-\infty}^{+\infty} \int_{-\pi/2}^{+\pi/2} W_x(s, n) \, W_h(t - s, \omega - n) \, \frac{dn}{2\pi},$$

which is a new form for the *uncertainty principle* of estimation evoked previously. This follows from the fact that, the smoothing function being a Wigner-Ville distribution (which is known to be support-preserving), the double bias of (5-11) cannot be reduced in one direction of the time-frequency plane without a corresponding increase in the other direction.

Considering now second order properties, we find (Flandrin and Martin, 1984):

$$(5-15) \quad \text{Cov}[S_x(t_1, \omega_1), S_x(t_2, \omega_2)] \approx$$

$$|A_h(\omega_2 - \omega_1, t_1 - t_2)|^2 \, W_x^2\left(\frac{t_1 + t_2}{2}, \frac{\omega_1 + \omega_2}{2}\right),$$

where

$$(5-16) \quad A_h(n, \tau) = \sum_{u=-\infty}^{+\infty} h(u + \tau) \, h^*(u - \tau) \, e^{inu}$$

is the symmetric *ambiguity function* (Sussman, 1962) of the observation window h(t). It follows from (5-13) that

$$(5-17) \quad \text{Var}[\,|S_x(t, \omega)|^2\,] \approx W_x^2(t, \omega).$$

Again, short-time periodograms, considered as estimators of the Wigner-Ville spectrum, furnish a double correlation which cannot be reduced in time (resp. frequency) without a corresponding increase in frequency (resp. time).

This choice results also in a high variability of the estimates, since the variance is of the order of the square of the true value, which is a natural extension of the stationary case studied in Welch, 1967. In a manner very similar to that considered in the case of Priestley's evolutionary spectra, this situation can be improved by replacing crude short-time periodograms by some smoothed versions in time like (5-7). It can then be shown that

$$(5-18) \quad \text{Var}[S'_x(t, \omega)] \approx \left|\sum_{s=-\infty}^{+\infty} g(s) \, A_h(0, s)\right|^2 W_x^2(t, \omega),$$

which reduces, in the case of a properly normalized rectangular window of length M, to (Flandrin and Martin, 1984)

(5-19) $\text{Var}[S'_x(t, \omega)] \approx \frac{1}{M} W^2_x(t, \omega)$.

Nevertheless, this variance reduction is obtained at the expense of a supplementary bias in time, leading necessarily to a severe bias-variance trade-off.

In fact, if we consider short-time periodograms, it appears that their limitations are due to the fact that their time and frequency behaviors are completely linked. The most simple way to overcome this difficulty is then to consider estimators characterized by a smoothing function which is *separable* in time and frequency, e.g.

(5-20) $\Pi(t, \omega) = g(t) W_h(0, \omega)$.

Inserting (5-20) into (5-10) comes up with the second example, called *smoothed pseudo-Wigner estimators* (Martin and Flandrin, 1983):

(5-21) $SPW_x(t, \omega) = 2 \sum\limits_{\tau = -\infty}^{+\infty} |h(\tau)|^2 [\sum\limits_{s = -\infty}^{+\infty} g(s) x(t + s + \tau) x^*(t + s - \tau)] e^{-i2\omega\tau}$

As expected, the double bias is now split up into two independent terms since

(5-22) $E[SPW_x(t, \omega)] = \int\limits_{-\pi/2}^{+\pi/2} [\sum\limits_{s = -\infty}^{+\infty} g(s - t) W_x(s, n)] W_h(0, \omega - n) \frac{dn}{2\pi}$.

This means that the time and frequency behaviors can now be controlled independently. As a consequence, bias in the time direction can be for instance totally removed if we do not apply any time smoothing, i.e. if g(t) is chosen as Dirac impulsive. We have then

(5-23) $E[SPW_x(t, \omega)] = \int\limits_{-\pi/2}^{-\pi/2} W_x(t, n) W_h(0, \omega - n) \frac{dn}{2\pi}$

and the bias concerns only the frequency direction. In a very general manner, we see that the time behavior is now controlled by the time smoothing function g(t), whereas the frequency behavior is controlled by the observation window h(t). This contrasts with the case of short-time periodograms for which the only observation window h(t) was controlling both behaviors.

The same applies to second order properties and correlation between estimates. We have especially a possibly total decorrelation for a sufficient spacing, namely

$$(5\text{-}24) \quad \text{Cov}[\text{SPW}_x(t_1, \omega_1), \text{SPW}_x(t_2, \omega_2)] \approx 0 \quad \begin{cases} |t_1 - t_2| > M \\ |\omega_1 - \omega_2| > 1/N \end{cases}$$

if M (resp. N) is the length of the smoothing window g(t) (resp. of the observation window h(t)).

Considering at last a (sufficiently large) rectangular window g(t), we obtain for the variance the approximate result (Flandrin and Martin, 1984)

$$(5\text{-}25) \quad \text{Var}[\text{SPW}_x(t, \omega)] \approx \frac{2}{M} W_x^2(t, \omega) ,$$

which is comparable to that obtained in the modified short-time periodograms case (5-19).

When using smoothed pseudo-Wigner estimators, the question is then only to adjust the smoothing function g(t) in order to achieve some optimum bias-variance trade-off in the time direction. An off-line solution has been proposed (Martin and Flandrin, 1983-1985a), which fits a piecewise stationary model to an observed process, frequency by frequency : the algorithm, which postulates the equivalence between Wigner-Ville spectra and power spectral densities on each stationary segment, finds an optimum segmentation by means of an Akaike-type criterion.

5.2.4. Two interpretation remarks

When defining estimation procedures, an implicit (but essential) question is : what are the estimators aimed at ? This may appear as trivial but, in fact, we have seen that the same quantity, namely the short-time periodograms, can be introduced for estimating either Priestley's evolutionary spectra or the Wigner-Ville spectrum, which are theoretically different. As a consequence, the performance of the estimators can only be derived w.r.t. the considered spectrum but, from a practical point of view, both are undistinguishable.

However, as far as slow evolutions of the processes are necessary for justifying the estimation procedures, we know that differences between both spectra are not dramatic (cf. Section 4.2.3.). This leads to the second remark according to which the necessary slow evolution assumptions also ensure a quasi-positivity of the Wigner-Ville spectrum to be estimated (cf. Section 4.6.). Therefore, we do possess with (4-19) and (5-10) a *coherent framework* for both the *definition* of a time-dependent spectrum (with almost all the desirable properties, including quasi-positivity) and its *estimation* (with a possible decoupling of the time and frequency properties).

6. Time-frequency formulation of some statistical Signal Processing problems

After having considered the definition and the estimation of time-dependent spectra, we will see in this Section that they can be used for natural, and physically meaningful, formulations of some classical Signal Processing problems.

6.1. Random time-variant filtering

As a first example, let us consider a channel with (time-variant) random impulse response $h(t, s)$. If $x(t)$ is a non-random input, the output $y(t)$ expresses as

$$(6-1) \quad y(t) = \int_{-\infty}^{+\infty} h(t, s) \, x(t - s) \, ds \, .$$

A common and important situation is the one for which

$$(6-2) \quad E[h(t_1, s_1) h^*(t_2, s_2)] = k(t_1 - t_2, s_1) \, \delta(s_1 - s_2) \, .$$

This corresponds to *wide-sense stationary uncorrelated scattering* (WSSUS) channels (Kennedy, 1969), such as those encountered e.g. in underwater acoustics.

Within this assumption, we obtain

$$(6\text{-}3) \quad E[\mathbf{W}_y(t, \omega)] = \iint_{-\infty}^{+\infty} K(t - s, \omega - n) \, W_x(s, n) \, ds \, \frac{dn}{2\pi} = C_x(t, \omega; K) ,$$

where

$$(6\text{-}4) \quad K(t, \omega) = \int_{-\infty}^{+\infty} k(t, s) \, e^{-i\omega s} \, ds$$

is the *scattering function* of the channel (Price, 1962).

In the formulation (6-3), the scattering function appears as a smoothing function tending to spread the Wigner-Ville distribution of the input, which is exactly its physical interpretation.

Furthermore, we know that, within the WSSUS assumption, the output power of the filter matched to the input but misaligned of (t_0, ω_0) expresses as (Green, 1962)

$$(6\text{-}5) \quad V(t_0, \omega_0) = \iint_{-\infty}^{+\infty} K(t, \omega) \, |A_x(\omega - \omega_0, t - t_0)|^2 \, dt \, \frac{d\omega}{2\pi} .$$

An equivalent formulation is then given by

$$(6\text{-}6) \quad V(t_0, \omega_0) = \iint_{-\infty}^{+\infty} C_x(t, \omega; K) \, W_x(t - t_0, \omega - \omega_0) \, dt \, \frac{d\omega}{2\pi} .$$

This can now be interpreted as a measure of the overlap between two modified Wigner-Ville distributions of the input : one smoothed by the scattering function, and another one shifted by the misalignement.

6.2. Optimum detection

As a second example, let us consider the classical binary decision problem

$$(6\text{-}7) \quad \left. \begin{array}{l} H_0 : r(t) = b(t) \\ H_1 : r(t) = b(t) + x(t) \end{array} \right\} t \in (T) ,$$

where $r(t)$ is the observation known on the time interval (T), $b(t)$ is stationary zero-mean white Gaussian noise of power spectral density N_0 and $x(t)$ a zero-mean Gaussian process to be detected.

We know (van Trees, 1971) that the classical formulation of the (maximum likelihood) optimum detector relies on the doubly orthogonal Karhunen-Loeve's expansion of the process

(6-8a) $x(t) = \sum_n x_n \phi_n(t)$

such that

(6-8b) $E[x_n x_m^*] = \lambda_n \delta_{nm}$; $\int_{(T)} \phi_n(t) \phi_m^*(t) \, dt = \delta_{nm}$,

where the λ_n and the $\phi_n(t)$ are respectively the eigenvalues and eigenfunctions of the covariance function $r_x(t, t')$ of the process $x(t)$:

(6-9) $\int_{(T)} r_x(t, s) \phi_n(s) \, ds = \lambda_n \phi_n(t)$, $t \in (T)$.

The detection is then achieved by thresholding the quantity

(6-10) $\Lambda = \sum_n \frac{\lambda_n}{\lambda_n + N_0} \left| \int_{(T)} r(t) \phi_n^*(t) \, dt \right|^2$.

Therefore, if we make use of *Moyal's formula* (Claasen and Mecklenbrauker, 1980a), which ensures a scalar product conservation :

(6-11) $\iint_{-\infty}^{+\infty} W_x(t, \omega) W_y(t, \omega) \, dt \, \frac{d\omega}{2\pi} = \left| \int_{-\infty}^{+\infty} x(t) y^*(t) \, dt \right|^2$,

it is immediate to replace (6-10) by the equivalent formulation

(6-12) $\Lambda = \iint_{(T)} W_r(t, \omega) \left[\sum_n \frac{\lambda_n}{\lambda_n + N_0} W_{\phi_n}(t, \omega) \right] dt \, \frac{d\omega}{2\pi}$.

This allows an explicit interpretation in terms of *correlation of time-frequency structures* (that of the observation and that of a reference), as described by Wigner-Ville distributions (Altes, 1986, Flandrin, 1988). It must be noticed that this interpretation, which may be thought of as intuitive for any time-frequency representation, is in fact valid only for those which satisfy Moyal's formula (6-11). This explains why time-frequency correlation procedures based e.g. on spectrograms (Altes, 1980) are suboptimum and need additional deconvolution procedures.

If we are now interested in *locally optimum detection* (Capon, 1961), i.e. optimum detection when the signal-to-noise ratio is small, all the eigenvalues are much smaller than the power spectral density of the additive noise and, as a first approximation, we get

$$(6\text{-}13) \quad \Lambda \approx \frac{1}{N_0} \iint\limits_{(T)} W_r(t, \omega) \, E[\mathbf{W}_x(t, \omega)] \, dt \, \frac{d\omega}{2\pi} .$$

since (Flandrin, 1982) :

$$(6\text{-}14) \quad E[\mathbf{W}_x(t, \omega)] = \sum_n \lambda_n \, W_{\phi_n}(t, \omega) .$$

The locally optimum receiver takes on a very simple form since it only consists in a comparison (in terms of correlation) between the Wigner-Ville distribution of the observation and the Wigner-Ville spectrum of the process to be detected, used as a reference (Flandrin, 1987b-1988).

As an illustration, we can consider the case of a non-random signal $x_d(t)$, jittered in both time and frequency, and affected by some Rayleigh fading factor. The resulting random process $x(t)$ can be modelled by

$$(6\text{-}15) \quad x(t) = b \, [\, x_d(t - t') \, e^{i\omega' t} \,] ,$$

where t' and ω' are random variables with a joint probability density function $G(t', \omega')$ and where b is a Gaussian random variable with unit variance. It can be immediately checked that

$$(6\text{-}16) \quad E[\mathbf{W}_x(t, \omega)] = C_{x_d}(t, \omega; G)$$

and

$$(6\text{-}17) \quad \Lambda \approx \frac{1}{N_0} \iint\limits_{(T)} W_r(t, \omega) \, C_{x_d}(t, \omega; G) \, dt \, \frac{d\omega}{2\pi} .$$

Therefore, the optimum receiver is clearly in accordance with what is suggested by intuition : detection is achieved by correlating the time-frequency structure of the observation with that of the reference, *smeared* by the probability density function (interpreted here as a smoothing function) of delay and frequency shift.

6.3. Time-frequency receivers and classifiers

If we consider (6-17) and symmetric functions G, we have in an equivalent way

$$(6-18) \quad \Lambda \approx \frac{1}{N_0} \iint_{(T)} W_{x_d}(t, \omega) \, C_r(t, \omega; G) \, dt \, \frac{d\omega}{2\pi}.$$

Relaxing then the interpretation of G as an *a priori* probability density function, and only retaining that of an *a posteriori* smoothing function, (6-18) offers a general class of time-frequency receivers with considerable flexibility (Flandrin, 1986a-1988). This is summarized in the following Table I where various receiver structures are presented as functions of some limiting cases for G

$G(t, \omega)$	$N_0 \Lambda$
$\delta(t) \, \delta(\omega)$	$\left\| \int_{(T)} r(t) \, x_d^*(t) \, dt \right\|^2$
$\delta(t)$	$\int_{(T)} \|r(t)\|^2 \, \|x_d(t)\|^2 \, dt$
$\delta(\omega)$	$\int_{-\infty}^{+\infty} \|R(\omega)\|^2 \, \|X_d(\omega)\|^2 \, \frac{d\omega}{2\pi}$
1	$\left[\int_{(T)} \|r(t)\|^2 \, dt \right] \left[\int_{(T)} \|x_d(t)\|^2 \, dt \right]$

Table I
Smoothing functions and associated receiver structures (6-18)

We conclude from this that very different receivers, such as *matched filters* (followed by an envelope detector), *energy detectors* or *intensity correlators* can be viewed as special cases of a unique formulation which moreover admits an intuitive and physically meaningful interpretation.

It must be noticed that the *two-dimensional correlation* which is involved in (6-18) can be simplified in the case of AM-FM signals for which

$$(6-19) \quad W_x(t, \omega) \approx 2\pi \, a_x^2(t) \, \delta(\omega - \omega_x(t)) \, .$$

The optimum procedure results then in a simple *path integration* along the instantaneous frequency curve (Kay and Boudreaux-Bartels, 1985; Flandrin, 1986a):

$$(6-20) \quad \Lambda \approx \frac{1}{N_0} \int_{(T)} a_x^2(t) \, C_r(t, \omega_x(t); G) \, dt \, .$$

By making use of generalized likelihood ratio tests, all these approaches can be extended to *estimation* problems (Kay and Boudreaux-Bartels, 1985; Flandrin, 1986a). Moreover, the same tools can be used for tackling *classification* of non-stationary signals directly in the time-frequency plane, for instance in failure detection problems (Chiollaz, Flandrin and Gache, 1987).

7. Conclusion

Considering its long history, time-dependent spectral analysis of non-stationary stochastic processes seems to raise never ending questions, and the proposed solutions, which have been developed here, are certainly not definitive answers to the addressed problem of a simultaneous time-frequency description. Nevertheless, it has been shown that, in a non-parametric context, looking for "natural" extensions to the stationary concept of power spectral density leads essentially to two main approaches : the first one is related to *Priestley's evolutionary spectra*, whereas the second one relies on the so-called *Wigner-Ville spectrum*. Owing both to objective considerations and personal experience, emphasis has been put on the second approach, providing hence a coherent framework for the *definition* and the *estimation* of a rather satisfactory time-dependent spectrum.

Space is too short for illustrating theory with concrete applications : these can be found in the literature and their quick development makes every day more difficult to review them. Apart from the numerous applications aimed at *deterministic* signals (see e.g. Boudreaux-Bartels, 1985 and the references therein), let us just mention for Priestley's evolutionary spectra : analysis of random vibrations (Hammond, 1973) or line tracking (Martin, 1981), and for the Wigner-Ville spectrum : analysis of biological time series (Jorres, Martin and Brinkmann, 1984) or fault detection in truck engines (Chiollaz, Flandrin and Gache, 1987), a more exhaustive list being reported e.g. in Flandrin, 1987a.

One great advantage of the methods discussed here is to be *non-parametric* in nature, and hence to need *no a priori model*. As a natural consequence, they are preffered candidates for a *blind* analysis, which can be of course understood as a first step, helping in the choice of an appropriate model, and preceding its fit.

More generally, time-frequency representations provide, when suitably defined, a natural complement to classical methods working either in the time domain or in the frequency domain. By means of the two dimensions which are involved in the time-frequency plane, a greater flexibility can be achieved, which allows satisfactory *descriptions* of non-stationary processes (for which time and frequency aspects are simultaneously relevant), and furthermore suggests to handle in a more efficient way their *processing*.

8. References

J.B. Allen, L.R. Rabiner, 1977
"A Unified Approach to Short-Time Fourier Analysis and Synthesis",
Proc. IEEE, **65** (11), pp. 1558-1564.

R.A. Altes, 1980
"Detection, Estimation and Classification with Spectrograms",
J. Acoust. Soc. Am., **67** (4), pp. 1232-1246.

R.A. Altes, 1986
"Some Theoretical Concepts for Echolocation",
NATO ASI on Animal Sonar Systems, Helsingør (DK).

F. Battaglia, 1979

"Some Extensions of the Evolutionary Spectral Analysis of a Stochastic Process",
Bull. Unione Matematica Italiana, **16-B** (5), pp. 1154-1166.

M.J. Bastiaans, 1981
"The Wigner Distribution Function of Partially Coherent Light",
Opt. Acta, **28** (9), pp. 1215-1224.

J.S. Bendat, A.G. Piersol, 1966
Measurement and Analysis of Random Data,
J. Wiley & Sons, New-York.

A. Blanc-Lapierre, B. Picinbono, 1955
"Remarques sur la Notion de Spectre Instantané de Puissance",
Publ. Sci. Univ. Alger B, **1**, pp. 2-32.

A. Blanc-Lapierre, B. Picinbono, 1981
Fonctions Aléatoires,
Masson, Paris.

G.F. Boudreaux-Bartels, 1985
"Time-Varying Signal Processing Using the Wigner Distribution Time-Frequency Signal Representation",
in : Advances in Geophysical Data Processing, vol. 2 (M. Simaan, ed.) pp. 33-79,
JAI Press, Greenwich.

N.G. de Bruijn, 1967
"Uncertainty Principles in Fourier Analysis",
in : Inequalities II (O. Shisha, ed.), pp. 57-71,
Academic Press, New-York.

J. Capon, 1961
"On the Asymptotic Efficiency of Locally Optimum Detectors",
IRE Trans. on Info. Theory, **IT-7**, pp. 67-71.

M. Chiollaz, P. Flandrin, N. Gache, 1987
"Utilisation de la Représentation de Wigner-Ville comme Outil de Diagnostic des Défauts de Fonctionnement de Moteurs Thermiques",
11eme Coll. GRETSI sur le Traitement du Signal et des Images, Nice (F), pp. 579-582.

T.A.C.M. Claasen, W.F.G. Mecklenbräuker, 1980a
'The Wigner Distribution - A Tool for Time Frequency Signal Analysis. Part
I : Continuous-Time Signals",
Philips J. Res., **35** (3), pp. 217-250.

T.A.C.M. Claasen, W.F.G. Mecklenbrauker, 1980b
"The Wigner Distribution - A Tool for Time Frequency Signal Analysis. Part
II : Discrete-Time Signals",
Philips J. Res., **35** (4/5), pp. 276-300.

T.A.C.M. Claasen, W.F.G. Mecklenbrauker, 1980c
"The Wigner Distribution - A Tool for Time Frequency Signal Analysis. Part
III : Relations with Other Time-Frequency Signal Transformations",
Philips J. Res., **35** (6), pp. 372-389.

L. Cohen, 1966
"Generalized Phase-Space Distribution Functions",
J. Math. Phys., **7** (5), pp. 781-786.

H. Cramer, 1961
"On Some Classes of Non-Stationary Stochastic Processes",
4th Berkeley Symp. on Math., Stat. and Proba., **2**, pp. 57-78,
Univ. Calif. Press.

H. Cramer, 1971
"Structural and Statistical Problems for a Class of Stochastic Processes",
The First S.S. Wilks Lecture at Princeton University, pp. 1-30,
Princeton University Press, Princeton,
reprinted in : <u>Random Processes : Multiplicity Theory and Canonical
Decompositions</u> (A. Ephremides and J.B. Thomas; eds.), pp. 305-334,
Dowden, Hutchinson & Ross, Inc., Stroudsburg.

P. Flandrin, 1982
"Representations des Signaux dans le Plan Temps-Fréquence",
These Doct. Ing., INPG. Grenoble (F).

P. Flandrin, 1986a
"On Detection-Estimation Procedures in the Time-Frequency Plane",
IEEE Int. Conf. on Acoust., Speech and Signal Proc. ICASSP-86, Tokyo (J), pp.
2331-2334.

P. Flandrin, 1986b
"On the Positivity of the Wigner-Ville Spectrum",
Signal Proc., **11** (2), pp. 187-189.

P. Flandrin, 1986c
"When is the Wigner-Ville Spectrum Non-Negative ?",
in : Signal Processing III : Theories and Applications (I.T. Young et al., eds.),
pp. 239-242,
North-Holland, Amsterdam.

P. Flandrin, 1987a
"Représentations Temps-Fréquence des Signaux Non-Stationnaires",
Thèse Doct. Etat ès Sc. Phys., INPG, Grenoble (F).

P. Flandrin, 1987b
"Détection Optimale Dans le Plan Temps-Fréquence",
11ème Coll. GRETSI sur le Traitement du Signal et des Images, Nice (F), pp.
77-80.

P. Flandrin, 1987c
"On the Spectrum of Some Fractional Brownian Motions",
IEEE Trans. on Info. Theory, to appear.

P. Flandrin, 1988
"A Time-Frequency Formulation of Optimum Detection",
IEEE Trans. on Acoust., Speech and Signal Proc., to appear.

P. Flandrin, B. Escudié, W. Martin, 1985
"Representations Temps-Fréquence et Causalité",
10ème Coll. GRETSI sur le Traitement du Signal et ses Applications, Nice (F),
pp. 65-70.

P. Flandrin, W. Martin, 1983
"Sur les Conditions Physiques Assurant l'Unicite de la Representation de
Wigner-Ville Comme Representation Temps-Frequence",
9ème Coll. GRETSI sur le Traitement du Signal et ses Applications, Nice (F),
pp. 43-49.

P. Flandrin, W. Martin, 1984

"A General Class of Estimators for the Wigner-Ville Spectrum of Non-Stationary Processes",
in : Lecture Notes in Control and Information Sciences 62 : Analysis and Optimization of Systems (A. Bensoussan, J.L. Lions, eds.), pp. 15-23, Springer-Verlag, Berlin.

O.D. Grace, 1981
"Instantaneous Power Spectra",
J. Acoust. Soc. Am., 69 (1), pp. 191-198.

P.E. Green, 1962
"Radar Measurements of Target Characteristics",
in : Radar Astronomy (J.V. Harrington, J.V. Evans, eds.),
Mc Graw-Hill, New-York.

Y. Grenier, 1983
"Time-Dependent ARMA Modeling of Non-Stationary Signals",
IEEE Trans. on Acoust., Speech and Signal Proc., ASSP-31 (4), pp. 899-911.

Y. Grenier, 1984
"Modélisation de Signaux Non-Stationnaires",
Thèse Doct. Etat ès Sc. Phys., Univ. Paris-Sud, Orsay (F).

Y. Grenier, 1987
"Parametric Time-Frequency Representations",
in : Traitement du Signal / Signal Processing (J.L. Lacoume, T.S. Durrani and R. Stora, eds.), Les Houches, Session XLV, pp. 339-397,
North-Holland, Amsterdam.

J.K. Hammond, 1973
"Evolutionary Spectra in Random Vibrations",
J. Roy. Stat. Soc. B, 35, pp. 167-188.

J.K. Hammond, R.F. Harrison, 1985
"Wigner-Ville and Evolutionary Spectra for Covariance Equivalent Non-Stationary Random Processes",
IEEE Int. Conf. on Acoust., Speech and Signal Proc. ICASSP-85, Tampa (FL), pp. 27.4.1-27.4.4.

R.L. Hudson, 1974
"When is the Wigner Quasi-Probability Density Non-Negative ?",

Rep. Math. Phys., **6** (2), pp. 249-252.

A.J.E.M. Janssen, 1988
"Positivity of Time-Frequency Distribution Functions",
Signal Proc., **14** (3), pp. 243-252.

R. Jorres, W. Martin, K. Brinkmann, 1984
"Identification of the Temperature Masking of the Circadian System of
Euglena Gracilis",
in : Cybernetics and Systems Research II (R. Trappl, ed.), pp. 293-297,
North-Holland, Amsterdam.

S. Kay, G.F. Boudreaux-Bartels, 1985
"On the Optimality of the Wigner Distribution for Detection",
IEEE Int. Conf. on Acoust., Speech and Signal Proc. ICASSP-85, Tampa (FL),
pp. 27.2.1-27.2.4.

R.S. Kennedy, 1969
Fading Dispersive Communication Channels,
J. Wiley & Sons, New-York.

LH. Koopmans, 1974
The Spectral Analysis of Time Series,
Academic Press, New-York.

M. Loève, 1962
Probability Theory,
D. van Nostrand Company, Princeton.

R.M. Loynes, 1968
"On the Concept of the Spectrum for Non-Stationary Processes",
J. Roy. Stat. Soc. B, **30** (1), pp. 1-20.

B.B. Mandelbrot, J.W. van Ness, 1968
"Fractional Brownian Motions, Fractional Noises and Applications",
SIAM Rev., **10** (4), pp. 422-437.

W.D. Mark, 1970
"Spectral Analysis of the Convolution and Filtering of Non-Stationary
Stochastic Processes",
J. Sound Vib., **11** (1), pp. 19-63.

W. Martin, 1981
"Line Tracking in Non-Stationary Processes",
Signal Proc., **3** (2), pp. 147-155.

W. Martin, 1982a
"Time-Frequency Analysis of Random Signals",
IEEE Int. Conf. on Acoust., Speech and Signal Proc. ICASSP-82, Paris (F), pp.
1325-1328.

W. Martin, 1982b
"Time-Frequency Analysis of Non-Stationary Processes",
IEEE Int. Symp. on Info. Theory ISIT-82, Les Arcs (F), p. 47.

W. Martin, 1984a
"Measuring the Degree of Non-Stationarity by Using the Wigner-Ville
Spectrum",
IEEE Int. Conf. on Acoust., Speech and Signal Proc. ICASSP-84, San Diego
(CA), pp. 41B.3.1-41B.3.4.

W. Martin, 1984b
"Wigner-Ville-Sprektralanalyse Nichtstationarer Prozesse - Theorie und
Anwendung in Biologischen Fragestellungen",
Habilitationsschrift, Univ. Bonn, Bonn (FRG).

W. Martin, P. Flandrin, 1983
"Analysis of Non-Stationary Processes : Short-Time Periodograms vs a
Pseudo-Wigner Estimator',
in : <u>Signal Processing II : Theories and Applications</u> (H.W. Schussler, ed.),
pp. 455-458,
North-Holland, Amsterdam.

W. Martin, P. Flandrin, 1985a
"Detection of Changes in Signal Structure by Using the Wigner-Ville
Spectrum",
Signal Proc., **8** (2), pp. 215-233.

W. Martin, P. Flandrin, 1985b
"Wigner-Ville Spectral Analysis of Non-Stationary Processes",
IEEE Trans. on Acoust., Speech and Signal Proc., **ASSP-33** (6), pp. 1461-
1470.

W. Martin, K. Kruger-Alef, 1986
"Application of the Wigner-Ville Spectrum to the Spectral Analysis of a
Class of Bio-Acoustical Signals Blurred by Noise",
Acustica, **61**, pp. 176-183.

G. Mélard, 1978
"Propriétés du Spectre Evolutif d'un Processus Non-Stationnaire",
Ann. Inst. H. Poincaré B, **XIV** (4), pp. 411-424.

C.H. Page, 1952
"Instantaneous Power Spectra",
J. Appl. Phys., **23** (1), pp. 103-106.

A. Papoulis, 1965
<u>Probability, Random Variables and Stochastic Processes</u>,
Mc Graw-Hill, New-York.

R. Price, 1962
"Detectors for Radar Astronomy",
in : <u>Radar Astronomy</u> (J.V. Harrington, J.V. Evans, eds.),
Mc Graw-Hill, New-York.

M.B. Priestley, 1965
"Evolutionary Spectra and Non-Stationary Processes",
J. Roy. Stat. Soc. B, **27** (2), pp. 204-237.

M.B. Priestley, 1981
<u>Spectral Analysis and Time Series</u>,
Academic Press, New-York.

A.W. Rihaczek, 1968
"Signal Energy Distribution in Time and Frequency",
IEEE Trans. on Info. Theory, **IT-14** (3), pp. 369-374.

R.A. Silverman. 1957
"Locally Stationary Random Processes",
IRE Trans. on Info. Theory, **IT-3**, pp. 182-187.

S.M. Sussman, 1962
"Least Squares Synthesis of Radar Ambiguity Functions",

IRE Trans. on Info. Theory, **IT-8**, pp. 246-254.

D. Tjøstheim, 1976
"Spectral Generating Operators for Non-Stationary Processes",
Adv. Appl. Prob., **8**, pp. 831-846.

H.L. van Trees, 1971
Detection, Estimation and Modulation Theory,
J. Wiley & Sons, New-York.

Y.H. Tsao, 1984
"Uncertainty Principle in Frequency-Time Methods",
J. Acoust. Soc. Am., **75** (5), pp. 1532-1540.

J. Ville, 1948
"Théorie et Applications de la Notion de Signal Analytique",
Cables et Transm., **2ème A** (1), pp. 61-74.

P.D. Welch, 1967
"The Use of Fast Fourier Transform for the Estimation of Power Spectra : a
Method Based on Time Averaging over Short, Modified Periodograms",
IEEE Trans. on Audio and Electroacoust., **AU-15**, pp. 70-73.

N. Wiener, 1930
"Generalized Harmonic Analysis",
Acta Math., **55**, pp. 117-258.

E.P. Wigner, 1932
"On the Quantum Correction for Thermodynamic Equilibrium",
Phys. Rev., **40**, pp. 749-759.

E.P. Wigner, 1971
"Quantum-Mechanical Distribution Functions Revisited",
in : Perspectives in Quantum Theory (W. Yourgrau, A. van der Merwe, eds.),
pp. 25-36,
MIT Press, Cambridge.

E. Wong, 1971
Stochastic Processes in Information and Dynamical Sciences,
Mc Graw-Hill, New-York.

PARAMETRIC TIME-FREQUENCY REPRESENTATIONS

Y. Grenier

Ecole Nationale Supérieure des Télécommunications, Paris, France

ABSTRACT.

This text describes parametric time-frequency representations, called here reliefs. After an analysis of the properties expected for such a relief, the classes of nonstationary signals are studied. One observe several equivalences between various definitions. This permits to define the class for which the concept of relief will be valid: the class of harmonizable nondegenerate signals. Two reliefs are presented in details, one defined by Priestley for oscillatory signals, the other defined by Tjøstheim using a commutation relation between two operators representing time and frequency. A third relief is also discussed: the rational relief. It is more adequate and simpler to use for ARMA signals. The estimation of these reliefs is also considered, through time-dependent ARMA modelling.

This chapter originally appeared in: J. L. Lacoume, T. S. Durrani and R. Stora, eds., Les Houches, Session XLV, 1985, Traitement du signal / Signal processing, © Elsevier Science Publishers B. V., 1987. Reprinted with permission.

1. INTRODUCTION.

It is almost a tradition to introduce the concept of time-frequency representation of a signal as was done by VILLE, 1948, BLANC-LAPIERRE, PICINBONO, 1955, PRIESTLEY, 1965 (and in the discussion of that paper G.A. BARNARD), DE BRUIJN, 1967, ESCUDIE, 1979, CLAASEN, MECKLENBRAUKER, 1980, GRACE, 1981, starting from music notation. A musical piece provides a good example of a nonstationary signal and since we perceive it as a sequence of temporal events, (i.e. the notes, to which we attribute a duration and a pitch, formalized in terms of frequency), it seems reasonable to infer that every signal should be or could be represented as the distribution of an energy in the two-dimensional space of time and frequency.

This function of time and frequency has received many names : instantaneous spectrum, evolutionary spectrum, joint time-frequency representation, time-frequency analysis..: here, a more condensed terminology will be prefered, and we will call it the relief of a signal, borrowing this term from geologists. After all, don't we talk of the crests and the valleys of a spectrum ?

Though the existence of such a function is almost evident, analysis of the relevant litterature shows that the problem of the definition of a relief does not admit a single solution. Therefore, before trying to answer the question "how to define...", it seems more judicious to ask first the question "why a relief..." In other words, we need to establish the conditions which a function of time and frequency has to satisfy before we can consider it to be a good representation. Two papers by BLANC-LAPIERRE, PICINBONO,1955, and by LOYNES, 1968, have investigated that topic. The object of section 2 is to summarize their approaches and their conclusions, as well as to comment on them.

The next sections are devoted to the definition of the relief in a parametric (stochastic) frame. This implies to delimit the class of nonstationary signals within which the definition of the relief is valid. It will be done in section 3. The difficulty is then to extend stationary spectral representations of random signals, without losing any desirable property. Two definitions belong to that framework : that given by PRIESTLEY, 1965, for a class of signals called "oscillatory", and that given by TJØSTHEIM, 1976-b, which is connected to Wold decomposition. These two approaches are described in section 4 and 5 respectively. The estimation aspects lead us to the analysis of the links between rational or ARMA models and reliefs. Section 6 describes the various models that can be associated with a rational time-varying impulse response, and section 7 deduces from that analysis, the definition of the rational relief, which is a variant of Tjøstheim's relief, while section 8 describes a class of estimators for this relief, based upon time-varying ARMA models.

2. REQUIRED PROPERTIES.

2.1 First approach.

BLANC-LAPIERRE, PICINBONO, 1955, emphasize the interpretation of the instantaneous power and the power spectral density of a signal as distributions of the energy. They also focus on the links between the representations of a signal and its filtered versions. They write three conditions C1 to C3, extending these properties to the concept of relief.

C1)   The margin laws of the relief $\rho(t, \omega)$ must reconstruct the signal power and spectrum.

C1-a)   The margin law in the frequency domain must be the power spectrum of the signal. This is expressed by (1) in the continuous time case.

(1)
$$\int_{-\infty}^{+\infty} \rho(t, \omega)dt = |Y(\omega)|^2$$

C1-b)   The margin law in time is the instantaneous power of the signal. This expressed by (2) in the continuous time case. The discrete time case is obtained simply from (1) by replacement of the integral by an infinite sum, and from (2) by replacement of the frequency integral by an integral over the interval $(-\pi, +\pi)$. This case will no longer be mentioned in this section.

(2)
$$\int_{-\infty}^{+\infty} \rho(t, \omega)\frac{d\omega}{2\pi} = |y(t)|^2$$

C1-c)   The integral of the relief is the total energy E (assumed finite):

(3)
$$\int_{-\infty}^{+\infty}\int_{-\infty}^{+\infty} \rho(t, \omega)dt\frac{d\omega}{2\pi} = E$$

C2)   Conditions connecting the relief of a filtered signal to the original one:

C2-a)   The relief of a signal filtered by a linear system with transfer function $G(\omega)$, should be obtained easily from $G(\omega)$ and the initial relief $\rho(t, \omega)$.

C2-b)   The relief $\rho(t, \omega)$ remains unchanged for all frequencies lower than ω_0, when the signal is low-pass filtered, with a cutoff frequency of ω_0:

(4)
$$Y(\omega) = \begin{cases} X(\omega) & \omega \leq \omega_0 \\ 0 & \omega > \omega_0 \end{cases} \Rightarrow \rho_y(t, \omega) = \begin{cases} \rho_x(t, \omega) & \omega \leq \omega_0 \\ 0 & \omega > \omega_0 \end{cases}$$

C2-c)   The relief remains unchanged up to time t_0 when the signal is cut at time t_0:

(5)
$$y(t) = \begin{cases} x(t) & t \leq t_0 \\ 0 & t > t_0 \end{cases} \Rightarrow \rho_y(t, \omega) = \begin{cases} \rho_x(t, \omega) & t \leq t_0 \\ 0 & t > t_0 \end{cases}$$

C3)   The relief of a signal should be positive.

   This set of conditions calls for some comments. The conditions C1 express the character of distribution of the relief in the time-frequency plane. A relief which also satisfies condition C3 can then be viewed as a power density: the quantity $\rho(t, \omega)dt.d\omega$ will be interpreted as the amount of energy located inside the rectangle $(t, t+dt).(\omega, \omega+d\omega)$.

The conditions C2-b and C2-c concern the locality of the relief. C2-b forces the relief to be causal, this seems natural and physical, since it allows the existence of real-time estimators of $\rho(t, \omega)$, i.e. estimators of $\rho(t, \omega)$ from the knowledge of y(t) up to t only. One may however doubt if it is necessary to require this in the framework of a definition which is a more theoretical than practical problem. Perhaps it would be preferable to express this locality in a noncausal way, for instance, by requiring that the first moment in t and in ω be equal to the group delay and the instantaneous frequency respectively.

Condition C2-a is justified by the fact that the power spectrum of a filtered signal is the product of the original spectrum by the squared modulus of the transfer function. Of course it would be nice to extend that property to the relief $\rho(t, \omega)$. However it can be proven easily that C2-a is satisfied as soon as the transformation signal $\rightarrow$ relief is invertible. This last condition seems to be more general.

2.2 Second approach.

Another set of conditions is given by LOYNES, 1968, with two variants numbered here as they are in the paper by LOYNES, 1968. The first variant is described by A1-A8.

A1)   The relief $\rho(t, \omega)$ must be real.

A2)   It will represent a distribution of the energy with respect to the frequency.

A3)   The relief of a signal linearly transformed (e.g. filtered) will be deduced from the original relief in a simple manner.

A4)   The relief $\rho(t, \omega)$ is a one-to-one mapping of the signal covariance.

A5)   If $x(t)$ is a stationary signal, its relief $\rho(t, \omega)$ is identical to the (ordinary) spectrum.

A6)   If a signal is composed of a succession of stationary parts, (e.g. $x(t)=x_1(t)$ if $t \leq O$ and $x(t)=x_2(t)$ if $t>0$ where $x_1(t)$ and $x_2(t)$ are two stationary signals), then the relief of $x(t)$ is also composed of the corresponding succession of stationary spectra.

$$(6) \qquad x(t) = \begin{cases} x_1(t) & t \leq 0 \\ x_2(t) & t>0 \end{cases} \Rightarrow \rho(t, \omega) = \begin{cases} X_1(\omega) & t \leq 0 \\ X_2(\omega) & t>0 \end{cases}$$

A7)   The estimation of $\rho(t, \omega)$ is possible at least in principle, from the knowledge of $x(t)$ on a finite interval.

A8)   The relief $\rho(t, \omega)$ is the Fourier transform of "some apparently meaningful quantity".

LOYNES, 1968, proposes another set of conditions B1-B12:

B1)   The relief is a real function $\rho(t, \omega)$ which is completely determined by the covariance function.

B2)   The relief $\rho(t, \omega)$ is a linear transform of the covariance function.

B3-7)   Same as A3-7.

B8)   The relief $\rho(t, \omega)$ is positive.

B9)   A modulation of the signal by a complex exponential with frequency ω_0 translates the relief $\rho(t, \omega)$ by a quantity ω_0 :

(7) $$y(t) = x(t).e^{j\omega_0 t} \implies \rho_y(t, \omega) = \rho_x(t, \omega + \omega_0)$$

B10)   The relief $\rho(t, \omega)$ of a shifted signal is obtained by the same shift of the original relief.

(8) $$y(t) = x(t+h) \implies \rho_y(t, \omega) = \rho_x(t+h, \omega)$$

B11)   The effects of time reversal and complex conjugation are the same on the signal and the relief :

$$y(t) = x^*(-t) \implies \rho_y(t, \omega) = \rho_x(-t, \omega)$$

$$y(t) = x(-t) \implies \rho_y(t, \omega) = \rho_x(-t, -\omega)$$

$$y(t) = x^*(t) \implies \rho_y(t, \omega) = \rho_x(t, -\omega)$$

$$x(t) \in R \implies \rho_x(t, \omega) = \rho_x(t, -\omega)$$

B12)   The relief $\rho(t, \omega)$ is a continuous function of the covariance of the signal.

2.3 Comparison between the two approaches.

It is interesting to gather conditions A1-A8 and B1-B12 by LOYNES, 1968 and conditions C1-C3 by BLANC-LAPIERRE, PICINBONO, 1955, since this shows clearly what is the minimal content of the concept of relief of a signal. It is possible to group the various conditions in four notions essentially, specifying the character of the relief as a distribution, its nature, its locality and its compatibility with linear operations applied to the signal, particularly filtering.

2.3.1 The relief as a distribution.

The fact that the relief should be a distribution of the energy of the signal over the time-frequency plane is expressed by conditions A2 and C1(a-b-c), involving integrals of the relief. One may also join to these conditions, A5/B5 requiring that for a stationary signal the relief is identical for every time t with the power spectrum of the signal, which will imply C1.

2.3.2 Nature of the relief.

The nature of the relief is specified in a multiple manner : it is required to be real (A1 and B1), with nonnegative values (B8 and C3). These are the properties ensuring that the relief is a power density, but one should add that it does not deny any physical meaning to a definition of the relief which would not satisfy this positive real condition. LOYNES,

1968, vacillates between four formulations of the links joining the relief to the covariance of a signal. The relief should be completely determined by the covariance (B1), it should be in a one-to-one correspondance with the covariance (A4/B4), it should be a linear transform of it (B2), or it should be a continuous function of it (B12). Together with condition A8 (the relief is the Fourier transform of a meaningful quantity) this leads to two conditions: the invertibility of the transform defining the relief, and the linearity of this transform, acting on the covariance through a Fourier transform.

2.3.3 Locality of the relief.

The third aspect of the relief is its locality expressed unformally by the condition: an event located at time t or inside an interval Δt containing t, does not influence the relief outside this interval, and by the analoguous condition on frequencies. Formally, this is expressed by the causality conditions C2-b and C2-c, but this contradicts the condition on the Fourier transform seen previously. The locality of the relief is described more strictly by condition A6/B6: the relief of a random signal which is composed of a succession of stationary slices is composed of the succession of spectra.

2.3.4 Invariance of the relief.

The last aspect is the invariance of the relief under several operations: filtering (C2-a and A3/B3, the relief of the filtered signal should be obtained from the original relief), frequency translation (B9, the relief should be translated in frequency by ω_0 when the signal is multiplied by a complex sinusoid with frequency ω_0), time shift (B10, the relief of a shifted signal should be obtained by a time shift of the original one), and finally, time reversal (B11).

2.4 The expected relief.

These three sets of conditions A, B and C, draw a portrait of the relief. This relief will be a positive real function of time and frequency, representing a distribution of the energy of the signal in the time-frequency plane. The margin laws or zero order moments will be the power envelope and the spectrum of the signal. The representation will be a quadratic function of the signal or a linear function of its covariance. It will satisfy locality requirements with respect to time or frequency or better, both. The relief will be invariant under several operations: linear filtering, time and frequency shift, time and frequency reversal.

The ideal definition of the relief being thus constrained, one may question the existence and unicity of this representation. Unfortunately, the answer is negative: there does not exist any definition simultaneously satisfying all the requirements. The two papers studied in this section do not provide a general demonstration of this fact, but they are right in attributing this negative result to the uncertainty relations, which forbids locating an event simultaneously in time and frequency, or measuring simultaneously the position (time-frequency) and the energy. LOYNES, 1968, gives a partial demonstration of the incompatibility between the reconstruction of the energy and power spectrum by the margin laws, and the linearity of the transformation from the covariance to the relief. Rigorous demonstrations of the impossibility of satisfying all the conditions at once, exist only in the nonparametric context. The conclusion of this section, is therefore that a general time-frequency representation of signals appears possible, but cannot be obtained unless we sacrifice at least one expected property.

3. CLASSES OF NONSTATIONARY SIGNALS.

Nonstationarity is only the absence of a property and is not sufficient to characterize the kind of signals which is considered. It is necessary to restrict the class of nonstationary signals in order to obtain spectral representations. This section describes several ways to define such a class, which finally melt into a single concept. The main results are obtained from the paper by MARTIN (W), 1982, from several references given there, and from the papers by MARTIN (M.M.), 1968, BOCHNER, 1956, GETOOR, 1956, NIEMI, 1976.

3.1 Harmonizable signals.

The property most commonly used in order to confine nonstationary signals within a practical class is harmonizability : the problem is to find a good compromise between stationary signals, which are severely constrained, but rich of properties, and the most general signals.

When one expects the possibility of defining a relief, one way to do that would be to start from spectral decompositions. Stationary signals provide the reference in terms of the doubly orthogonal decomposition on the basis of complex exponentials.

$$(9) \qquad\qquad y(t) = \int\limits_{-\infty}^{+\infty} e^{j\omega t} dY(\omega)$$

The first orthogonality is that of the functions, the complex exponentials which are orthogonal with respect to the scalar product of time functions :

$$(10) \qquad\qquad <e^{j\omega_1 t}, e^{j\omega_2 t}> = \int\limits_{-\infty}^{+\infty} e^{j(\omega_1 - \omega_2)t} dt = \delta(\omega_1 - \omega_2)$$

The second orthogonality is that of the increments of $Y(\omega)$ which are orthogonal random variables :

$$(11) \qquad\qquad E(dY(\omega_1).dY^*(\omega_2)) = \delta(\omega_1 - \omega_2).dS(\omega_1).d\omega_2$$

$\delta(\omega)$ being Dirac distribution.

When the signal $y(t)$ is no longer stationary, if it is harmonizable, the definition of harmonizability shows that it is still possible to write (9), but the increments $dY(\omega)$ are no longer orthogonal. If $\Phi(\omega_1, \omega_2)$ is the bidimensional Fourier transform of the covariance R(t,s) of $y(t)$, the covariances of the increments $dY(\omega)$ are given by (12).

$$(12) \qquad\qquad E(dY(\omega_1).dY^*(\omega_2)) = d\Phi(\omega_1, \omega_2)$$

One can view the problem posed by the nonstationarity as due to the double orthogonality which cannot be satisfied in (12). In order to satisfy it, one has to replace the complex exponentials by the eigenfunctions of the covariance operator, according to

Karhunen's theorem. This is valid for finite energy signals, but was also extended to finite power signals (BLANC-LAPIERRE, 1979, BLANC-LAPIERRE, PICINBONO, 1981). The orthogonality becomes true for the scalar product of time functions, and that defined by the covariance. Unfortunately, without the complex exponentials, the concept of spectrum disappears so that this representation cannot define a relief.

Another remark is that this class of harmonizable signals is too large for a correct definition of the relief. MARTIN, 1982, has shown that a degeneracy condition was also necessary. With this assumption, one obtains a class which had been described by several authors, in various forms which were then shown to be equivalent.

3.2 UBLS signals, stationary dilatations, propagators.

The starting point for the definition of that class of signals is the question of the existence of the shifts operators. Given a signal $y(t)$, we would like to define the shifts operators U_t, such that for any time s, we get $y(s+t)=U_t y(s)$. It was shown by GETOOR, 1956, that this is possible for signals having the following nondegeneracy property:

Definition (stationary nondegeneracy):
A signal $y(t)$ with covariance $R(t,s)$ is said to be stationarily nondegenerate if and only if the existence of n instants t_i and n complex numbers a_i such that

$$\sum_{i=1}^{n} \sum_{j=1}^{n} a_i a_j^* R(t_i, t_j) = 0$$

implies for any $s > 0$ that:

$$\sum_{i=1}^{n} \sum_{j=1}^{n} a_i a_j^* R(t_i+s, t_j+s) = 0$$

Theorem (GETOOR, 1956):
If the covariance $R(t,s)$ of the signal $y(t)$ satisfies the property of stationary nondegeneracy, then the shifts U_t exist.

Proof:
The shifts are defined over H, the closure of the linear space of all combinations $\sum a_i y(t_i)$ by $U_t y(s) = y(s+t)$ for any s and t. We simply need to show that $y=0$ implies $U_t y=0$ for any t. Any null vector $y=0$ is a combination of some $y(t_i)$, then:

$$y = \sum a_i y(t_i) = 0 \Rightarrow |y|^2 = 0 = \sum a_i a_j^* R(t_i, t_j) \Rightarrow |U_t y|^2 = 0 \Rightarrow U_t y = 0$$

Therefore the definition of U_t is correct. ∎

The shifts being defined, we can examine what are their properties. If the signal was stationary, the shifts would be unitary. In the nonstationary context, which was considered here, we cannot force such a property, but it would be nice if the shifts were bounded. A sufficient condition for that is given by:

Theorem (GETOOR, 1956):
If for every s, there exists a constant M_s such that for every choice of instants $t_1, \dots t_n$ and

of complex numbers $a_1, \ldots a_n$ one has:

(13)
$$\sum_{i=1}^{n} \sum_{j=1}^{n} a_i a_j^* R\,(t_i+s,t_j+s) \le M_s \sum_{i=1}^{n} \sum_{j=1}^{n} a_i a_j^* R\,(t_i,t_j)$$

then the shifts U_t are bounded in $H\,(y)$.

Proof:
If property (13) is satisfied, the signal is stationarily nondegenerate, then the shifts exist, and M_s is a bound for $|U_s|$. ∎

If we want the shifts to be uniformly bounded, we have to drop out the index s in M_s. This last condition is the source of the definition of uniformly bounded linearly stationary (UBLS) signals given by TJØSTHEIM, THOMAS, 1975, and extended to the strict case by TJØSTHEIM, 1976-a.

Definition:
A signal $y\,(t)$ is said to be UBLS if and only if there exists a positive constant M such that for every choice $t_1, \ldots t_n, a_1, \ldots a_n$, and for every s, one has:

(14)
$$E\left|\sum_{i=1}^{n} a_i y\,(t_i+s)\right|^2 \le M.E\left|\sum_{i=1}^{n} a_i y\,(t_i)\right|^2$$

It is clear that (14) is only rewriting (13), when all the M_s are bounded. This class of UBLS signals is the same as the class which MARTIN, 1968-b, called the (weakly) almost periodic signals, defined by the two conditions: 1) that there exist a group of shift operators U_t for $y\,(t)$, such that $U_t y\,(s)=y\,(t+s)$, and 2) that this group be uniformly bounded, i.e. that there exists a real $c>0$ such that $|U_t|<c$ for all t. By the way, this implies the condition $0<\dfrac{1}{c} \le |U_t| \le c$ for all t.

We can then come to another concept: that of rational similitude, or stationary dilatations, for a signal $y\,(t)$, i.e the existence of a bounded operator B and a stationary signal $x\,(t)$ such that $y\,(t)=Bx\,(t)$. The name of rational similitude was given by NIEMI, 1976, but the concept had been formulated by MARTIN, 1968-b, using a theorem by SZ-NAGY, 1947, on bounded operators. The use of Sz-Nagy's theorem (see appendix) ensures that, to the group of linear uniformly bounded operators U_t, one can associate a bounded operator with bounded inverse B such that $B^{-1}U_t B$ is unitary. Then for every $x\,(0)$, the signal $x\,(t)=B^{-1}U_t Bx\,(0)$ is stationary. If $x\,(0)=B^{-1}y\,(0)$, then $x\,(t)$ is given by: $x\,(t)=B^{-1}y\,(t)$ or $y\,(t)=Bx\,(t)$ so that there exists a rational similitude for $y\,(t)$. Conversely, if $y\,(t)$ possesses a rational similitude, it is uniformly bounded linearly stationary, this is a special case of a theorem by TJØSTHEIM, THOMAS, 1975, which states that if $x\,(t)$ is uniformly bounded linearly stationary and if B is a bounded operator with bounded inverse, $y\,(t)=Bx\,(t)$ is also uniformly bounded linearly stationary (a stationary signal is of course UBLS with $M=1$). If $y\,(t)$ is UBLS and has a mean spectrum, that is, if the limit

$$\lim_{t \to \infty} \frac{1}{T} \int_0^T <y(t+\tau),y(t)> = S(\tau)$$

exists, then NIEMI, 1976, shows that there exists for $y(t)$ a rational similitude $x(t)$ with correlation $S(\tau)$.

Since these demonstrations of the existence of a stationary dilatation are rather abstract, a restricted example can give some insight into this concept. Consider the discrete-time signal y_t, and its Wold decomposition (CRAMER, 1961-a, b):

$$y_t = \sum_{i=0}^{\infty} h_i(t)\varepsilon_{t-i}$$

Assume that any time-function $h_i(t)$ in this decomposition admits a Fourier transform:

$$h_i(t) = \int_{-\pi}^{+\pi} H_i(\omega)e^{j\omega t}d\omega$$

Then define the signals $x_\omega(t)$ as:

$$x_\omega(t) = \sum_{i=0}^{\infty} H_i(\omega)\varepsilon_{t-i}$$

The signals $x_\omega(t)$ are stationary. The signal y_t is then written as:

$$y_t = \int_{-\pi}^{+\pi} x_\omega(t)e^{j\omega t}d\omega$$

which can also be written as $y_t = B.x_t$, with B the Fourier transform (bounded operator, due to Parceval's relations), and x_t, a stationary signal taking its values within the Hilbert space of (generalized) functions over $]-\pi,+\pi[$: x_t is a stationary dilatation for y_t.

Another characterization of this class of signals is obtained by replacing the shift group U_t by the semi-group of propagators $U(t,s)$, such that $y(t)=U(t,s)y(s)$. They have to verify the three properties (MARTIN, 1982):

1) $U(t,t) = I,$

2) $U(t,s) = U(t,r).U(r,s),$

3) $U(t,s)$ is continuous with respect to t and s.

MASANI, 1978, has shown in a very abstract and large framework that the existence of such propagators for a signal was equivalent to the property of being UBLS. More precisely, theorem 3.4 by MASANI, 1978, states that $y(t)$ has a propagator if and only if $y(t)$ satisfies one of the two conditions (13) or (14) where M is replaced by M_s. Moreover, the theorem shows that the best possible choice for M_s is the square of the norm of $U(t+s,t)$.

3.3 Definition of the maximal class.

Starting from a demonstration given by Howland, MARTIN, 1982, has shown that the existences of a propagator and of a rational similitude were equivalent, but this requires an additionnal hypothesis: $y(t)$ has the property of stationary nondegeneracy. Using the rational similitude $y(t)=Px(t)$, and the shift operator V_t of $x(t)$, which is unitary, the propagator is introduced by $U(t,s)P=PV_{t-s}$, and its properties are easily verified. The reciprocal is more difficult and requires the definition of operators V_s such that $V_s f(t)=U(t,t-s)f(t-s)$, acting on $K=L^2(R,H(y))$. They form a group of bounded operators, to which Sz-Nagy's theorem associates a bounded operator B with bounded inverse such that $\tilde{V}_s=B^{-1}V_sB$ is unitary. It is then shown that $\tilde{V}_s$ is a shift operator, and a construction of $x(t)$ follows.

We have now for the signal $y(t)$ four equivalent properties:

1)   to be uniformly bounded linearly stationary,

2)   to possess a rational similitude,

3)   to possess a group of bounded shifts operators,

4)   to possess a propagator $U(t,s)$.

As a consequence, $y(t)$ also satisfies the property of stationary nondegeneracy: a degeneracy cannot be local, if the variance of $y(t)$ vanishes at $t=t_0$, it vanishes for every t.

The next step is to connect this class with harmonizable signals, and more precisely with the class of harmonizable nondegenerate signals (as called by MARTIN, 1982). It is clear that if $y(t)$ possesses a rational similitude, it is harmonizable. The converse is shown by MIAMEE, SALEHI, 1978. They have also shown the equivalence with the class of V-bounded signals, which had been introduced by BOCHNER, 1956, containing the signals such that for every choice of instants $t_1, ...t_n$ and complex numbers $a_1, ...a_n$:

$$(15) \qquad E\left|\sum_{i=1}^{n} a_i y(t_i)\right| \leq M.E\left(\underset{\omega}{Sup}\left|\sum_{i=1}^{n} a_i e^{-j\omega t_i}\right|\right)$$

This condition can also be expressed as:

$$(16) \qquad \left|\int_{-\infty}^{+\infty}\phi(-t)y(t)dt\right| \leq M.\underset{\omega}{Sup}\,|\Phi(\omega)|$$

for every function $\phi(t)$ with summable modulus, and Fourier transform $\Phi(\omega)$. BOCHNER, 1956, had shown that every harmonizable signal was V-bounded, and the converse was shown by MIAMEE, SALEHI, 1978, while NIEMI, 1976, showed that if $y(t)$ was uniformly bounded linearly stationary,it was V-bounded. One has to add to the four preceding properties:

5)   to be harmonizable nondegenerate,

6)   to be V-bounded and nondegenerate.

This set of 6 equivalent properties constitutes the definition of the maximal class of nonstationary signals.

4. RELIEF DEFINED BY PRIESTLEY.

4.1 Definition of the relief by Priestley.

PRIESTLEY, 1965, tries to find a compromise between the decomposition in terms of complex exponentials, which preserves the concept of frequency, but does not ensure orthogonality, and the doubly orthogonal decomposition which loses the concept of frequency. Assume that a spectral representation of the covariance, following Karhunen's theorem, is given by (17):

$$(17) \qquad R(t,s) = \int \phi_t(\omega)\phi_s^*(\omega)d\mu(\omega)$$

where $\mu(\omega)$ is a positive measure. The associated spectral representation of $y(t)$ is:

$$(18) \qquad y(t) = \int \phi_t(\omega)dY(\omega)$$

with:

$$(19) \qquad E(dY(\omega_1).dY^*(\omega_2)) = \delta(\omega_1-\omega_2)d\mu(\omega_1)$$

PRIESTLEY, 1965, introduces a class of signals called "oscillatory", characterized by the special form of $\phi_t(\omega)$:

$$(20) \qquad \phi_t(\omega) = A_t(\omega)e^{j\theta(\omega)t}$$

In this nonunique decomposition of $\phi_t(\omega)$ (ϕ itself being nonunique), the time function $A_t(\omega)$ is constrained to be low frequency, i.e. its Fourier transform has a maximum at frequency 0. The function $\theta(\omega)$ represents the frequency of $\phi_t(\omega)$, but a change of variables $\omega \rightarrow \omega'$ ensures that $\theta(\omega) = \omega'$, so that (20) becomes:

$$(21) \qquad \phi_t(\omega) = A_t(\omega)e^{j\omega t}$$

The decomposition (18) with functions as in (21) has orthogonal increments $dY(\omega)$, and is "almost orthogonal" with respect to the scalar product of time functions $\phi_t(\omega)$, in the sense that these are products of complex exponentials (which are orthogonal) by slowly varying functions $A_t(\omega)$. This leads PRIESTLEY, 1965, to the definition of the relief $\rho(t, \omega)$ by (22):

$$(22) \qquad \rho(t,\omega)d\omega = |A_t(\omega)|^2 d\mu(\omega)$$

Rewriting decomposition (18), gives:

$$(23) \qquad y(t) = \int e^{j\omega t}A_t(\omega)dY(\omega)$$

so, $y(t)$ has an interpretation as a combination of complex exponentials, with amplitudes $A_t(\omega)dY(\omega)$, and (22) has the physical meaning of a spectral distribution at time t.

A second interpretation of (23) justifies the presence of this definition among parametric representations. If we assume that $A_t(\omega)$ has a continuous (inverse, $\omega \rightarrow u$) Fourier transform $h_t(u)$:

$$(24) \qquad A_t(\omega) = \int_{-\infty}^{+\infty} e^{-j\omega u} h_t(u) du$$

we can incorporate this expression in (23) and define a signal $s(t)$ by (25). This signal is stationary since the increments $dY(\omega)$ are orthogonal.

$$(25) \qquad s(t) = \int_{-\infty}^{+\infty} e^{j\omega t} dY(\omega)$$

we obtain:

$$(26) \qquad y(t) = \int_{-\infty}^{+\infty} h_t(u) s(t-u) du$$

and in (26) $y(t)$ can be intrepreted as the output of a time-varying linear system with impulse response $h_t(u)$, the input being a stationary signal $s(t)$: this is clearly a parametric representation.

4.2 Comments on Priestley's relief.

The definition given by PRIESTLEY, 1965, was soon criticized because of the weakness of some of its properties. MELARD, 1978-b, questioned the choice of the class "oscillatory" signals, since it is very difficult to prove that a signal belongs to this class, for instance autoregressive moving-average (ARMA) signals are not necessarily oscillatory, despite their strong structure as outputs of systems with rational impulse responses. This class has other drawbacks, DE SCHUTTER-HERTELEER, 1979, showed that it could not be extended properly to vector signals, because the relief which comes out implies time-invariant coherence between any two components, and this has no physical meaning. BATTAGLIA, 1977, showed that the class of oscillatory signals is not closed with respect to addition, the sum of two oscillatory signals is itself oscillatory if and only if:

$$\frac{\partial}{\partial t} \left[\frac{\rho_1(t, \omega)}{\rho_2(t, \omega)} \right] = 0$$

He then introduced the class of σ-oscillatory signals, which are finite sums of oscillatory signals, but this generalization does not eliminate other drawbacks. If an oscillatory signal is filtered, the result is not always oscillatory, there are some conditions on the filter: its transfer function has to be maximal at frequency 0 (MM. MARTIN, 1968-a, RAO, TONG, 1974)

Because of these restrictions, the definition of the relief given by PRIESTLEY, 1965, despite its intuitive aspect and its immediate physical interpretation, is not often employed (there were several uses for time-series analysis in econometrics and also in biology, MARTIN, 1981).

4.3 Wold decomposition and Priestley's relief.

Wold decomposition represents a discrete-time signal y_t as an infinite order MA signal, i.e. the output of a system with time-varying infinite impulse response, the input being a white noise:

$$(27) \qquad\qquad y_t = \sum_{s=-\infty}^{t} h(t,s)\varepsilon_s$$

In the continuous-time case, $\varepsilon(t)$ will be a brownian motion and $y(t)$ will be written (for a signal with multiplicity 1):

$$(28) \qquad\qquad y(t) = \int_{-\infty}^{t} h(t,s)d\varepsilon(s)$$

The class of oscillatory signals introduced by PRIESTLEY, 1965, is not disconnected to Wold decomposition since ABDRABBO, PRIESTLEY, 1967, have given conditions under which an oscillatory signal admits this decomposition. If the relief is:

$$(29) \qquad\qquad \rho(t,\omega) = |A_t(\omega)|^2 f(\omega) \quad \text{with} \quad f(\omega)d\omega = d\mu(\omega)$$

then, the conditions in the discrete-time case are:

1) $\qquad \int_{-\pi}^{+\pi} Log\, f(\omega)d\omega > -\infty$

2) $\qquad A_t(\omega)$ for every time t is the Fourier transform of a causal sequence $g_t(u)$, $u \geq 0$.

3) $\qquad \int_{-\pi}^{+\pi} Log\, |A_t(\omega)|^2 d\omega > -\infty$

4) $\qquad \int_{-\pi}^{+\pi} Log\, \rho(t,\omega) > -\infty$

This last condition is satisfied as soon as 1 and 3 are; 2 is also a consequence of 3. In the continuous-time case, conditions 1, 3 are sufficient and become:

1) $\qquad \int_{-\infty}^{+\infty} \dfrac{Log\, f(\omega)}{1+\omega^2} d\omega > -\infty$

2) $\qquad \int_{-\infty}^{+\infty} \dfrac{Log\, |A_t(\omega)|^2}{1+\omega^2} d\omega > -\infty$

These conditions appear clearly as nonstationary extensions of the classical conditions for a stationary signal to be purely nondeterministic (or to admit a decomposition like (27) or (28)).

5. RELIEF DEFINED BY TJØSTHEIM AND MELARD.

We now study the definition of the relief that was given by Tjøstheim and Mélard. It would be simpler to follow MELARD, 1978-a, which defines the relief directly from the Wold decomposition of the signal $y(t)$, as it was extended to the nonstationary case by CRAMER, 1961-a, b. Unfortunately the definition obtained by this approach is rather arbitrary. It is better to follow TJØSTHEIM, 1976-b, who shows explicitly in which sense his definition is canonical. The price to pay for that is a slight increase in the complexity of the demonstration, using some basic results from operator theory (see appendix). This approach is chosen here, because it gives much more insight into the meaning of the definition. Like other authors, TJØSTHEIM, 1976-b, tries to extend a property from the stationary case to the nonstationary one, but unlike the others, he does not extend a spectral representation like (13), but a commutation relation between two operators, one associated to time and the other to frequency.

This commutation relation comes from quantum mechanics, where the measurements of the position of a particle and its momentum are mean values of two operators, q for the position and p for the momentum. The particle is represented by its wave function $\phi \in L^2$; it has two representations on two particular basis : $\phi(t)$ and its Fourier transform, $\Phi(\omega)$. In these basis the action of operators p and q is simply :

$$(30) \qquad \left[p\phi\right](t) = -j\frac{d\phi(t)}{dt} \quad \text{and} \quad \left[p\Phi\right](\omega) = \omega d\Phi(\omega)$$

$$(31) \qquad \left[q\phi\right](t) = tf(t) \quad \text{and} \quad \left[q\Phi\right](\omega) = j\frac{d\Phi(\omega)}{d\omega}$$

It is easy to show that the operators satisfy the commutation relation :

$$(32) \qquad pq - qp = -jI$$

Any pair of operators satisfying relation (32) will be said to be a Schrödinger couple. Equations (30)-(32) are a direct consequence of the properties of the Fourier transform. For random signals, the same commutation relation was established by TJØSTHEIM, 1976-c, for two operators T and H : T will be the time operator and H the energy operator or Hamiltonian of the stationary signal $y(t)$. The next paragraphs summarize the following steps of the approach by TJØSTHEIM, 1976-a and b : 1) introduction of operators T and H, 2) demonstration of the commutation relation in continuous time, then in discrete time, and 3) definition of spectral generators in discrete time, and in continuous time.

5.1 Operators T and H.

To the random signal $y(t)$, one associates $L(y)$, the linear space spanned by all the combinations $\sum a_i y(t_i)$, and the Hilbert space $H(y)$, closure of $L(y)$ for the scalar product given by $<x_1,x_2>=E(x_1 x_2^*)$, the signal being assumed to be complex valued. One defines the growing sequence of spaces H_t (CRAMER, 1961-a) spanned by the combinations $\sum a_i y(t_i)$ for $t_i \leq t$. The space $H_{-\infty}$ will be their intersection : $H_{-\infty} = \cap_t H_t$. The signal $y(t)$ is assumed to be purely nondeterministic, which is expressed by $H_{-\infty}=0$. The sequence of spaces H_{t+0} is constructed from H_t by $H_{t+0} = \cap_n H_{t+\frac{1}{n}}$. One introduces the operator P_t of projection over H_{t+0}. The set of projectors P_t is a resolution of the identity (see appendix) which defines a self-adjoint operator by:

$$(33) \qquad\qquad T = \int t . dP_t$$

The operator T is the time operator, while the energy operator H is simply the infinitesimal generator of the continuous group of unitary operators composed of the family of shifts U_s introduced in section 3. The resolution of the identity $E(\omega)$ for the operators U_s is given by (A-7) and it implies for H:

$$(34) \qquad\qquad H = \int \omega . dE(\omega) \quad \text{with } U_s = e^{jHs}$$

The following derivation explains why H is called Hamiltonian:

$$\frac{d}{dt} y(t) = \frac{d}{dt} U_t y(0) = \frac{d}{dt} \left[e^{jHt} \right] y(0) = jH . e^{jHt} y(0) = jH y(t)$$

This last equation is similar to Schrödinger's equations in quantum mechanics, where H is the Hamiltonian of the system. H is, of course, self-adjoint, since (34) is its spectral decomposition. Conversely one may associate to the self-adjoint operator T the group of unitary operators S_t such that $S_t = e^{jTt}$.

5.2 Commutation relation.

The demonstration of the commutation relation $TH - HT = jI$ contains two steps: a commutation between the projectors P_t and the shifts U_s: $P_t U_s = U_s P_{t-s}$, and a commutation between the shifts U_s and the operators S_t known as Weyl's relations. A theorem by Von Neumann then permits one to deduce the relation between T and H.

Proof:

demonstration of $P_t U_s = U_s P_{t-s}$:

$$\forall x(t): y = P_{t-s} x(t) \Leftrightarrow <y - x(t), x(u)> = 0, \ u \leq t-s$$

$$\Leftrightarrow <U_s y - U_s x(t), U_s x(u)> = 0, \ u \leq t-s$$

$$\Leftrightarrow <U_s y - U_s x(t), x(v)> = 0, \ v \leq t$$

$$\Leftrightarrow U_s y = P_t U_s x(t)$$

hence

$$U_s P_{t-s} x(t) = P_t U_s x(t) \text{ for } every \ x(t).$$

demonstration of Weyl's relations (TJØSTHEIM, 1976-a):

$$S_t U_s = \int e^{j\lambda t} dP_\lambda U_s = \int e^{j\lambda t} U_s dP_{\lambda - s}$$

$$S_t U_s = U_s \int e^{j\lambda t} dP_{\lambda - s} = U_s e^{jst} \int e^{j(\lambda - s)t} dP_{\lambda - s}$$

$$S_t U_s = e^{jst} U_s S_t$$

demonstration of the commutation relation (BEALS, 1971, pp 68-69):

$$S_t U_s = e^{jTt} s^{jHs} = e^{jst} U_s S_t = e^{jst} e^{jHs} e^{jTt}$$

by formal derivation with respect to t:

$$jT.e^{jTt} e^{jHs} = js.e^{jst} e^{jHs} e^{jTt} + e^{jst} e^{jHs} jT.e^{jTt}$$

for $t = 0$ one obtains:

$$jT.e^{jHs} = jse^{jHs} + e^{jHs} jT$$

by formal derivation with respect to s:

$$j^2 THe^{jHs} = j^2 s^2 He^{jHs} + je^{jHs} + j^2 He^{jHs} T$$

for $s = 0$ one obtains:

$$j^2 TH = jI + j^2 HT$$

so: $HT - TH = jI$, the commutation relation sought. ∎

In discrete time, the commutation relation will be different: the time operator T remains unchanged, but H, the infinitesimal generator of the shifts, is replaced by the shift U of one time unit. The relation is $TU - UT = U$. Its demonstration requires the introduction of the innovation $\varepsilon_t = y_t - P_{t-1} y_t$. The operator T is defined by $T = \sum_t t (P_t - P_{t-1})$.

From that we deduce succesively the relations $T\varepsilon_t = t\varepsilon_t$, $P_t U = UP_{t-1}$, $U\varepsilon_t = \varepsilon_{t-1}$ and $TU - UT = U$.

Demonstration:
$T\varepsilon_s = s\varepsilon_s$ is shown by a simple computation:

$$T\varepsilon_s = \sum_t t\,(P_t - P_{t-1})(y_s - P_{s-1}y_s)$$

$$= \sum_t t\,(P_t y_s - P_{t-1}y_s - P_{min\,(t,s-1)}y_s + P_{min\,(t-1,s-1)}y_s)$$

$$= \sum_{t=-\infty}^{s-1} t\,(P_t y_s - P_{t-1}y_s - P_t y_s + P_{t-1}y_s)$$

$$+s\,(P_s y_s - P_{s-1}y_s)$$

$$+ \sum_{t=s+1}^{\infty} t\,(y_s - y_s - P_{s-1}y_s + P_{s-1}y_s)$$

hence $T\varepsilon_s = s\,(y_s - P_{s-1}y_s) = s\varepsilon_s$.

The relation $P_t U = U P_{t-1}$ is obtained exactly as in the continuous time case, and the following one is easy to derive:

$$U\varepsilon_s = U\,(y_s - P_{s-1}y_s)$$

$$= Uy_s - UP_{s-1}y_s$$

$$= Uy_s - P_s Uy_s$$

$$= y_{s+1} - P_s y_{s+1}$$

$$= \varepsilon_{s+1}$$

The commutation relation follows:

$$(TU - UT)\varepsilon_s = TU\varepsilon_s - UT\varepsilon_s$$

$$= T\varepsilon_{s+1} - U\,(s\varepsilon_s)$$

$$= (s+1)\varepsilon_{s+1} - s\varepsilon_{s+1}$$

$$= \varepsilon_{s+1}$$

$$= U\varepsilon_s$$

This relation being satisfied for every ε_s, and these ε_s spanning $H\,(y)$, the relation is proved. ■

Now we have shown that to every stationary random signal $y\,(t)$, one can associate a Schrödinger couple of self-adjoint operators: T, a time operator connected to the time decomposition (Wold) of $y\,(t)$, and H, an energy operator connected to the spectral representation of $y(t)$. Since, for nonstationary signals, the set of spaces $H(y,t)$ and the family of projectors P_t is still well defined (CRAMER, 1961-a, b), while the shift, from which H is deduced, is no longer unitary, TJØSTHEIM, 1976-b, proposes to define H through the commutation relations. An operator H (resp. U) such that the couple (T,H) (resp. (T,U)) is a Schrödinger couple will be said "spectral generating operator". It remains to prove the existence and unicity of such a spectral generating operator.

5.3 Spectral generating operators (discrete time) and relief.

To show the existence in the discrete time case of a spectral generating operator U, or an operator such that $TU-UT=U$, TJØSTHEIM, 1976-b, uses the normalized innovation η_t. This trick can be simply understood from the preceding demonstration of the commutation relation. If the signal $y(t)$ is nonstationary, the time operator T remains defined and still has the property $T\varepsilon_s=s\varepsilon_s$ so, $T\eta_s=s\eta_s$ where $\eta_s=E(|\varepsilon_s|^2)^{-\frac{1}{2}}\varepsilon_s$. If it is possible to introduce an operator U such that $U\eta_s=\eta_{s+1}$, the second and third parts of the demonstrations are valid. This operator U is easily constructed since η_s, the normalized innovation is a stationary white noise with unit variance, and admits a unitary shift U. Unfortunately, this operator U is not unique : if η_t is replaced by $\varepsilon_t^{(\alpha)}=e^{j\alpha(s)}\varepsilon_t$, the operator U_α defined by $U_\alpha\varepsilon_t^{(\alpha)}=\varepsilon_{t+1}^{(\alpha)}$ is still a spectral generating operator. However, TJØSTHEIM, 1976-b, shows that every spectral generating operator can be written as U_α for a certain sequence $\alpha(t)$. He then requires the operator U_α to be canonical in the sense that it must reduce to a shift operator in the stationary case : $\alpha(s)$ is necessarily a constant with arbitrary value since this value does not change the properties of U_α; this permits the choice $\alpha(s)=0$.

The spectral generating operator provides a spectral representation of y_t close to (18), which is canonical in the sense of the commutation relation, while the representation proposed by PRIESTLEY, 1965, needed auxiliary conditions, which did not even make it unique. Let us write the Wold decomposition y_t as given by CRAMER, 1961-a, b :

$$(35) \qquad y_t = \sum_{s=-\infty}^{t} h(t,s)\varepsilon_s$$

when the innovation is normalized, defining $\gamma(t,s)=h(t,s)g(s)^{\frac{1}{2}}$ with $g(s)=E(|\varepsilon_s|^2)$, we obtain the time representation :

$$(36) \qquad y_t = \sum_{s=-\infty}^{t} \gamma(t,s)\eta_s$$

The decomposition according to (A-6) of the operator U with simple spectrum and generator η_0 leads to :

$$(37) \qquad \eta_t = U^t\eta_0 = \int_{-\pi}^{+\pi} e^{j\omega t} dE(\omega)\eta_0$$

and defining $dE(\omega)\eta_0=d\mu(\omega)$, to :

$$(38) \qquad y_t = \int_{-\pi}^{+\pi} F(t,\omega)d\mu(\omega) \quad \text{and} \quad F(t,\omega) = \sum_{s=-\infty}^{t} \gamma(t,s)e^{j\omega s}$$

The quantity $F(t,\omega)$ permits us to define a relief $\rho(t,\omega)$ (and MELARD, 1978-a proceeds this way):

$$(39) \qquad \rho(t,\omega) = |F(t,\omega)|^2 = \left| \sum_{s=-\infty}^{t} \gamma(t,s)e^{j\omega s} \right|^2$$

Notice that this relief is defined if and only if it is possible to normalise the innovation; this implies that its variance remains greater than a quantity strictly positive and this is exactly the nondegeneracy condition in section 3. The class of nonstationary signals where the definition of the relief given by TJØSTHEIM, 1976-b, and MELARD, 1978-a is valid is precisely the class of harmonizable nondegenerate signals.

5.4 Spectral generating operators (continuous time) and relief.

For continuous time signals $y(t)$, the difficulty comes first from the operator T defined by (33) which has no longer a simple spectrum. If its multiplicity is M, and if $\{Z_i, i=1..M\}$ is a set of generators for T, the Wold decomposition (CRAMER, 1964 and 1966) is:

$$(40) \qquad y(t) = \sum_{i=1}^{M} \int_{-\infty}^{t} h_i(t,s)d\varepsilon_i(s) \quad \text{and} \quad d\varepsilon_i(s) = dP_s Z_i$$

In the particular case where $M=1$, assuming that $y(t)$ has stable innovations (nondegenerate), TJØSTHEIM, 1976-b, constructs a spectral generating operator H just as in the discrete time case, normalizing the innovation, and deriving a continuous group of unitary operators U_s defined by $U_s\varepsilon(\Delta)=\varepsilon(\Delta+s)$ where ε is the normalized innovation, Δ a set with finite measure and $\Delta+t$ the same set after a shift of time t. Reproducing the demonstration of the commutation relation, but using the properties of orthogonality of the innovation ε, it is possible to show the equality $P_t U_s = U_s P_{t-s}$, and the rest of the demonstration is valid, the commutation relation follows for T and H, the infinitesimal generator of the group U_s. The unicity of the operator H raises the same problems as in the discrete time case. The spectral representation is deduced from the Wold decomposition, or through the isomorphism between $H(y)$ and L^2, which associates the Schrödinger couple (H,T) with the couple (p,q) in (30) and (31):

$$(41) \qquad y(t) = \int_{-\infty}^{+\infty} F(t,\omega)d\Phi(\omega)$$

$$(42) \qquad \Phi(\omega_1)-\Phi(\omega_2) = \frac{1}{\sqrt{2\pi}} \int_{-\infty}^{+\infty} \frac{e^{-j\omega_2 s}-e^{-j\omega_1 s}}{-js}d\varepsilon(s)$$

$$(43) \qquad F(t,\omega) = \int_{-\infty}^{+\infty} h(t,s)e^{j\omega s}ds$$

In these equations, $\varepsilon(s)$ is the normalized innovation and $h(t,s)$ comes from the Wold decomposition:

$$(44) \qquad y(t) = \int_{-\infty}^{t} h(t,s)d\varepsilon(s)$$

This decomposition, as well as its discrete time counterpart (38) was established by KOREZLIOGLU, 1963, but he did not apply it to the definition of a relief $\rho(t,\omega)$. TJØSTHEIM, 1976-b, did it:

$$(45) \qquad \rho(t,\omega) = |F(t,\omega)|^2 = \left| \int_{-\infty}^{+\infty} h(t,s)e^{j\omega s}ds \right|^2$$

This definition has two weakness, first, it is not so clearly canonical as it was in the discrete time case, second, for a signal $y(t)$ with multiplicity M greater than 1, it leads TJØSTHEIM, 1976-b, to define the relief as an M-dimensional vector, its i-th component $\rho_i(t,\omega)$ being the relief deduced from the partial innovation $\varepsilon_i(t)$, according to (44) and (45). Such a vector spectral representation for a scalar signal seems surprising. TJØSTHEIM, 1976-b, considers that it is not more annoying than the multiplicity of the time representation. However, it is possible to go further. If one writes:

$$(46) \qquad y(t) = \sum_{i=1}^{M} \int_{-\infty}^{+\infty} F_i(t,\omega)d\Phi_i(\omega)$$

with $\Phi_i(\omega)$ defined as in (42), where $\varepsilon(s)$ is taken as the normalized partial innovation $\varepsilon_i(s)$, and $F_i(t,\omega)$ defined as in (43) where h is also indexed by i (see (40) normalized), then, one writes the variance of $y(t)$:

$$(47) \qquad E(|y(t)|^2) = \sum_{i=1}^{M} \sum_{j=1}^{M} \int\int_{-\infty}^{+\infty} F_i(t,\omega)F_j^*(t,\xi)d\Phi_i(\omega)d\Phi_j^*(\xi)$$

The spectral measures $\Phi_i(\omega)$ are twice orthogonal: each one as an orthogonal measure: $E(d\Phi_i(\omega)d\Phi_i^*(\xi))=\delta(\omega-\xi)d\omega/2\pi$ and together, by the orthogonality of the subspaces $H(z_i,t)$ for every t. Therefore (47) can be written:

$$(48) \qquad E(|y(t)|^2) = \int_{-\infty}^{+\infty} \left[\sum_{i=1}^{M} |F_i(t,\omega)|^2 \right] \frac{d\omega}{2\pi}$$

This last equation shows that one may define the relief by:

$$(49) \qquad \rho(t,\omega) = \sum_{i=1}^{M} |F_i(t,\omega)|^2$$

$$= \sum_{i=1}^{M} \left| \int_{-\infty}^{+\infty} h_i(t,s)e^{j\omega s}ds \right|^2$$

To go more deeply into this problem, the choice between these two formulations for the relief by Tjøstheim, in the continuous time case (vector or norm of this vector) is not necessary if one is rectricted to the class of harmonizable nondegenerate signals seen in section 3 : every signal with bounded shift is the projection of a stationary signal (MARTIN, 1968-b) and therefore, has a multiplicity equal to 1. The relation (45) constitutes a complete definition of the relief in this class of signals.

5.5 Properties of the relief by Tjøstheim and Mélard.

What are the properties of the relief defined by Tjøstheim-Mélard, in the discrete time case by (39) and in the continuous time case by (45)? MELARD, 1978-a, has checked for this relief the set of properties proposed by LOYNES, 1968. According to the conclusion of section 2, it appears that this relief, since it satisfies the positivity condition, and since it cannot simultaneously satisfy all the required properties, must have some defects from the point of view of locality. One notices immediately, from (48), that the relief defined by Tjøstheim and Mélard has some of the required locality properties, that numbered C1-b in section 2 : the marginal distribution in time is the variance of the signal. The marginal distribution in frequency cannot be considered because the signal does not have a finite energy. One should rather consider the mean distribution computed as the limit as T tends to infinity of the mean over the interval $(-T, +T)$. But to which mean spectrum should this limit be compared? In fact, the only property which can be checked is the identity of the relief of a stationary signal with its power spectral density, and the definition given by Tjøstheim-Mélard, satisfies this property by construction. In the sense settled in section 2, this relief is a correct distribution of the energy of the signal in the time-frequency plane.

The nature of this relief is characterized by its reality.(A1, B1), and its positivity (B8, C3). But the links with the covariance of the signal asked by LOYNES, 1968, (B1, B2, A4, B4, B12) are not established by this definition as MELARD, 1978-a, showed through a counterexample. Given a white noise ε_t, let us generate a signal y_t as in (50) :

$$(50) \qquad y_t = \begin{cases} \varepsilon_t - 0.5\varepsilon_{t-1} & \text{if } t \neq 1 \\ 0.5\varepsilon_1 - \varepsilon_0 & \text{if } t = 1 \end{cases}$$

This signal has the same relief as the stationary signal $y_t = \varepsilon_t - 0.5\varepsilon_{t-1}$, while their covariances are different, therefore, there is not a one-to-one mapping between covariance and relief (A4, B4).

The third aspect of the relief studied in section 2 is its locality. The causality property expressed by BLANC-LAPIERRE, PICINBONO, 1955, (condition C2-b) is also satisfied by construction of this relief. Its frequency equivalent cannot be established, but the condition of the succession of slices (A6/B6), has been analysed deeply by MELARD, 1978-a, who showed that the commutation (say at time 0) of a stationary signal $y_{1,t}$ to another stationary signal $y_{2,t}$, implies the commutation of the relief at the same instant from the spectrum $Y_1(\omega)$ to the spectrum $Y_2(\omega)$ provided y_1 and y_2 have the same innovation. At the opposite extreme, if their innovations are independent, then the relief is equal to $Y_1(\omega)$ up to time 0, and evolves after time 0 towards $Y_2(\omega)$, but reaches $Y_2(\omega)$ only after an infinite time. The locality by slices is not completely satisfied by this definition.

The last aspect is the invariance of the relief. It is immediately verified that the invariance under a time shift (B9) or a frequency shift (B10) is satisfied. The invariance under filtering is only approximately satisfied: MELARD, 1978-a, shows that if the filter has a finite impulse response and if the signal y_t has a relief $\rho(t,\omega)$, the variations of which remain bounded by a quantity $B(\omega)$ independent of time, over any interval shorter than the duration of the filter response, then the relief of the filtered signal can be reconstructed approximately from the original relief. The property of invariance under time reversal (B11) cannot be satisfied, for the mere reason that the innovation is not invariant under this transformation: the function $f(t,s)$ is not invariant, nor is the relief.

5.6 Connections with the z transform.

It is interesting to seek the connection between Tjøstheim's definition of the relief and the various approaches towards the notion of a nonstationary z transform. Such a generalization was sought by ZADEH, 1961, who defined $H(t,z)$ as the response of the system to an input $e^{j\omega t}$, divided by $e^{j\omega t}$. JURY, 1964, showed that $H(t,z)$ can be deduced from the impulse response $h(t,s)$, by:

$$(51) \qquad H(t,z) = \sum_{i=0}^{\infty} h(t,t-i)z^{-i}$$

The squared modulus of this time-varying impulse response $\rho(t,\omega)=|H(t,e^{j\omega})|^2$ is exactly the relief defined by Tjøstheim, to which it gives a nice physical interpretation. This is however a very partial interpretation since this extension of the z-transform shares very few properties with the classical (time-invariant) definition. Extension of the z-transform with a better behavior have been given recently by KAMEN, KHARGONEKAR, 1982.

5.7 Comments on Tjøstheim's and Mélard's relief.

The definition of the relief done by TJØSTHEIM, 1976-b, faces the compromise between positivity and locality inherent to Heisenberg's fourth uncertainty relation by keeping the positivity and neglecting the condition of locality within slices. This choice has two consequences which have a minor influence from a theoretical point of view, but are more annoying practically. The first consequence is that it is possible to construct simple examples of autoregressive signals for which the evolution of the relief follows very poorly the evolution of the model. The gravity of this situation is increased by the major importance of autoregressive models in many applications. Let us assume that the signal y_t is synthesized from two signals $y_{1,t}$ and $y_{2,t}$ both autoregressive and stationary, and for the sake of simplification, of the same order p:

$$(52) \qquad y_{k,t} + a_{k,1} y_{k,t-1} + ... + a_{k,p} y_{k,t-p} = \varepsilon_{k,t} \quad (k=1,2)$$

The signal y_t is obtained by a commutation from one signal to the other:

(53)
$$\begin{cases} y_t = y_{1,t} & \text{if } t \leq 0 \\ y_t = y_{2,t} & \text{if } t \geq 1 \end{cases}$$

MELARD, 1978-a, shows that if $\varepsilon_{1,t}$ and $\varepsilon_{2,t}$ are independent, then the relief of y_t is identical to $Y_1(\omega)$, the spectrum of $y_{1,t}$, up to time 0, and it evolves towards $Y_2(\omega)$ which it reaches after an infinite time. However, if we consider the signal y_t for $t > 0$, and its innovation, we can prove that after p samples, the innovation, as well as the parameters of the autoregressive model, become identical to what they are for the signal $y_{2,t}$:

(54) $\varepsilon_t = \varepsilon_{2,t}$ and $a_i(t) = a_{2,i}$ for $t > p$

Proof:

The innovation ε_t is the difference between y_t and its estimate given its past:

$$\varepsilon_t = y_t - E(y_t / y_{t-1}, y_{t-2}, ...)$$

for $t > 0$, since $\varepsilon_{1,t}$ and $\varepsilon_{2,t}$ (or $y_{1,t}$ and $y_{2,t}$) are independent:

$$\varepsilon_t = y_t - E(y_t / y_{t-1}, y_{t-2}, ..., y_1)$$

$$\varepsilon_t = y_{2,t} - E(y_{2,t} / y_{2,t-1}, y_{2,t-2}, ..., y_{2,1})$$

the fact that $y_{2,t}$ is autoregressive with order p implies:

$$\varepsilon_t = y_{2,t} - E(y_{2,t} / y_{2,t-1}, ..., y_{2,t-p}) \text{ for } t > p$$

hence:

$$\varepsilon_t = \varepsilon_{2,t} \text{ for } t > p \text{ so } a_i(t) = a_{2,i} \text{ for } t > p.$$

We may also add that the autoregressive order in the interval $1 \leq t \leq p$, is equal to $t-1$, and that the model at time t is the order $t-1$ model in the Levinson recursion leading to the model $(a_{2,1}, ... a_{2,p})$ of order p. ∎

Because of this behavior of the autoregressive model, one could expect that the relief, if it was sufficiently local or instantaneous, would follow $Y_1(\omega)$ until $t=0$, then have a transition of duration p, and finally be equal to $Y_2(\omega)$ from $t=p+1$ to ∞. As we have seen, the definition given by Tjøstheim and Mélard does not behave in this manner.

One could make the following objection: this example sounds quite artificial, this kind of commutation from one signal, y_1, to another, y_2, totally independent from the previous one, is rarely observed. Let us make the opposite hypothesis, the two signals have the same innovation (which is also the innovation of y_t) $\varepsilon_t = \varepsilon_{1,t} = \varepsilon_{2,t}$. In this case, the autoregressive order for y_t remains constant, and the commutation from model 1 to model 2 is done instantaneously when passing from sample $t=0$ to sample $t=1$. However the relief in the sense given by Tjøstheim-Mélard still has the same behavior. Since the impulse response of the system generating y_t from its innovation ε_t, is equal to model 1 for $t \leq 0$, but evolves slowly and asymptoticaly towards the impulse response of model 2, for positive times, so does the relief. The reason for the weakness of Tjøstheim's and Mélard's relief with respect to the condition of locality by slices clearly appears to lie in the use of the impulse response of the system generating the signal, for this representation is not a local one.

The second weakness mentioned earlier, is connected to a practical situation where this requirement of locality by slices appears again, namely in the estimation of the relief. A signal being known only by a realization on a finite interval of time, $(0,T)$, the estimation of Tjøstheim's and Mélard's relief is impossible without an auxiliary hypothesis on the signal before the first measured sample: null signal, stationary signal with the same generating system as the one at time 0, any compromise between these two hypotheses... We face the paradox that even if a model for the signal, for instance autoregressive or ARMA, is known, this is not yet sufficient to determine the relief. These considerations have drawn me to the conclusion that the relief by Tjøstheim-Mélard could be enhanced in the direction of a better locality, if the generator model was a more local representation than the impulse response. This leads to a rational relief (GRENIER, 1981-a and b), which is less general than the relief by Tjøstheim-Mélard, but more practical. Several authors had made this replacement before, but their use of an ARMA model within the definition was not consistent. It will be shown in section 7 that the correct definition requires a state-space model.

6. CLASSES OF NONSTATIONARY MODELS.

6.1 Rational nonstationary models.

Two kinds of models are good candidates for the replacement of the impulse response, ARMA models and state-space representations. In the stationary case , the ARMA model is defined by the order p of its denominator (more precisely of the denominator of its transfer function), by the order q of its numerator, and by the coefficients $a_1 \cdots a_p$ and $b_0...b_q$ of the recurrence relation between the normalized input ε_t (its variance is equal to 1), and the output y_t :

$$(55) \qquad y_t + a_1 y_{t-1} + ... + a_p y_{t-p} = b_0 \varepsilon_t + b_1 \varepsilon_{t-1} + ... + b_q \varepsilon_{t-q}$$

In the nonstationary case, the coefficients a_i and b_i become functions of time, and the conditions of stability and minimum phase become conditions of absolute summability of the impulse responses of the filter and its inverse (HALLIN, MELARD 1977, MILLER, 1969, HALLIN, 1978). Two extensions of (55) are possible. The first one seems more natural :

$$(56) \qquad \sum_{i=0}^{p} a_i(t) y_{t-i} = \sum_{j=0}^{q} b_j(t) \varepsilon_{t-j} \quad \text{and} \quad a_0(t) = 1$$

But it is also possible to select the time indexes in a_i to be t-i, so that the parameters and the variable which they multiply, have the same index. This will be called the "synchronous" form, the first one being called the "shifted" form.

$$(57) \qquad \sum_{i=0}^{p} a_i(t-i) y_{t-i} = \sum_{j=0}^{q} b_j(t-j) \varepsilon_{t-j}$$

Of course, passing from one form to the other is merely a problem of notation, and it is trivial to derive the coefficients in one form from the coefficients of the other one. If we write $\alpha_i(t)$ for the shifted functions, we have $\alpha_i(t)=a_i(t-i)$ or $a_i(t)=\alpha(t+i)$. Nonetheless, this discussion is not a peculiar one, when defining the relief, since the intuitive idea is to freeze the coefficients of the model at time t, compute the spectrum associated to this model and decide that this constitutes the relief $\rho(t,\omega)$ at time t. The synchronous and shifted versions of the ARMA model will not give the same relief, and the difference between the two reliefs may be dramatic, for instance in the case of periodic autoregressions.

6.2 Impulse response and ARMA model.

The equivalence between ARMA models and impulse responses is well known in the stationary case: an ARMA(p,q) model is equivalent to an $AR(\infty)$ model or an $MA(\infty)$ model, this last model giving the Wold decomposition of y_t. The numerator, the denominator, and the model itself have transfer functions:

$$(58) \qquad A(z) = 1+a_1z^{-1}+...+a_pz^{-p}$$

$$(59) \qquad B(z) = b_0+b_1z^{-1}+...+b_qz^{-q}$$

$$(60) \qquad H(z) = \frac{B(z)}{A(z)}$$

This last function H(z), computed by long division of B by A (each one being a polynomial in z^{-1}) is just the z transform of the filter impulse response or the equivalent $MA(\infty)$ model. The $AR(\infty)$ model is obtained by long division of A by B. In the nonstationary case, this procedure cannot be used, since the z-transform does not apply. Everything has to be determined in the time domain. Here, we consider only the association of an impulse response, or an $MA(\infty)$ model, to a finite order ARMA model.

Proposition :
The nonstationary system with impulse response $h(t,s)$ admits a rational or ARMA representation as in (57) if and only if there exist two integers p, q, and p functions $a_i(t)$, i=1...p such that:

$$(61) \qquad \sum_{i=0}^{p} a_i(t-i)h(t-i,t-n) = 0 \text{ for } n > q$$

where $a_0(t) = 1$.

Proof :
Necessary condition :
Assume that the system given by $h(t,s)$ is ARMA(p,q), as in (57), use :

$$y_t = \sum_{k=0}^{\infty} h(t,t-k)\varepsilon_{t-k}$$

where ε_t is any input, (57) is transformed into:

(62)
$$\sum_{i=0}^{p} a_i(t-i) \sum_{k=0}^{\infty} h(t-i,t-k-i)\varepsilon_{t-k-i} = \sum_{j=0}^{q} b_j(t-j)\varepsilon_{t-j}$$

The two members of this equation must be equal for any sequence ε_t, this implies relation (61) for $n > q$, taking $j = k+i$ in (62) and ensures the necessary condition.

Sufficient condition:

Assume that there exist p,q and the $a_i(t)$ satisfying (61) and define:

(63)
$$b_j(t) = \sum_{i=0}^{j} a_i(t+j-i) h(t+j-i,t)$$

Then the system with impulse response $h(t,s)$ satisfies the relation (62), and therefore, the recurrence relation (57), which proves the sufficiency. ∎

Corollary:

The signal y_t having a Wold decomposition with impulse response $h(t,s)$ is an ARMA(p,q) signal if and only if there exist two integers p, q and p functions $a_i(t)$ satisfying condition (61).

Proof:

The demonstration is short. The preceding proposition ensures that y_t is the output of a system defined by the recurrence equation (57), and $h(t,s)$ being the impulse response in the Wold decomposition of y_t is absolutely summable as well as the response of the inverse system. Since $h(t,s)$ is also the impulse response associated to (57), it follows that (57) is the correct definition of an ARMA model, and ε_t is the innovation of y_t. ∎

Remark:

We have seen how to replace the impulse response $h(t,s)$ by the recurrence relation (ARMA). The inverse is easy and the impulse response of the ARMA model is:

(64)
$$h(t,t) = b_0(t)$$

(65)
$$h(t,t-k) = b_k(t-k) - \sum_{i=1}^{k} a_i(t-i) h(t-i,t-k) \quad k \in (1,q)$$

(66)
$$h(t,t-k) = - \sum_{i=1}^{min(p,k)} a_i(t-i) h(t-i,t-k) \quad k > p$$

6.2.1 Canonical observable form.

Beside the ARMA representation, we have to consider the state variable representation. It can be obtained under two different canonical forms: the controllable form and the observable form. The next paragraphs describe how to obtain them from the ARMA model. The observable form does not raise any difficulty, but the controllable form requires more care. These two canonical forms are well known in the stationary case, but were almost never studied in the nonstationary case, the only approaches being the algebraic approach given by KAMEN, HAFEZ, 1979, and the recent results by BITTANTI, BOZERN, GUARDABASSI, 1985. The following propositions (GRENIER, 1984), are a partial response to the question of existence and construction of these canonical forms, only in the context of scalar signals or single-input, single-output systems.

Proposition:
The system described by the recurrent ARMA equation (57) admits the state-space representation (67) and reciprocally:

$$
(67) \qquad x_t = \begin{bmatrix} -a_1(t-1) & 1 & 0 & 0 \\ \cdot & & 0 & 1 & 0 \\ \cdot & & & & 1 \\ -a_n(t-1) & 0 & & 0 \end{bmatrix} x_{t-1} + \begin{bmatrix} b_0(t) \\ \vdots \\ b_{n-1}(t) \end{bmatrix} \varepsilon_t
$$

$$
y_t = [\ 1 \quad 0 \ . \ . \ 0] x_t
$$

Demonstration:
This is done by direct computation, defining the state x_t :

$$
x_t(1) = y_t
$$

$$
x_t(2) = y_{t+1} + a_1(t) y_t - b_0(t+1) \varepsilon_{t+1}
$$

$$
x_t(3) = y_{t+2} + a_2(t) y_t - b_0(t+2) \varepsilon_{t+2} - b_1(t+1) \varepsilon_{t+1}
$$

$$
x_t(n) = \sum_{i=0}^{n-1} a_i(t+n-i-1) y_{t+n-i-1} - \sum_{i=0}^{n-2} b_i(t+n-i-1) \varepsilon_{t+n-i-1}
$$

It is then clear that (57) implies (67) and reciprocally with $n = sup\,(p, q+1)$. ■

Corollary:
The system with impulse response $h(t,s)$ admits the representation (67) if and only if there exist two integers p, q and p functions $a_i(t)$ such that (61) is satisfied. The coefficients $b_i(t)$ are then given by (63).

6.2.2 Canonical controllable form.

Proposition:
The system with impulse response $h(t,s)$ admits the state-space representation :

$$(68) \quad x_t = \begin{bmatrix} 0 & 1 & 0 & & 0 \\ & 0 & 1 & & \\ & & & & 0 \\ 0 & & 0 & & 1 \\ -a_n(t-n) & . & . & -a_1(t-1) \end{bmatrix} x_{t-1} + \begin{bmatrix} 0 \\ \\ 0 \\ 1 \end{bmatrix} \varepsilon_t$$

$$y_t = \begin{bmatrix} b_{n-1}(t) & . & . & b_0(t) \end{bmatrix} x_t$$

if and only if there exist two integers p and q, and p functions of time $a_i(t)$ such that:

$$(69) \quad \sum_{i=0}^{p} a_i(t-m)h(t,t-m+i) = 0 \quad \text{for } m > q$$

then $n = \sup(p, q+1)$ and the coefficients $b_i(t)$ are given by:

$$(70) \quad b_j(t) = \sum_{i=0}^{j} a_i(t-j)h(t,t-j+i) \quad \text{for } j=0..n-1$$

Proof:

Necessary condition:
If we write (68) under the form
$$\begin{cases} x_t = A_{t-1}x_{t-1}+Be_t \\ y_t = C_t x_t \end{cases}$$
and we define the state transition matrix $\Phi(t,s)$, the impulse response becomes:
$$h(t,t-m) = C_t\Phi(t,t-m)G$$

with $\Phi(t,t-m)=A_{t-1}\cdots A_{t-m}$. Let $V_k(t,t-m)$ be the k-th column of the matrix $\Phi(t,t-m)$. A recurrence relation on these columns is a consequence of the properties of $\Phi(t,t-m)$ and of A_t:

$$\Phi(t,t-m+1).A_{t-m} = \Phi(t,t-m) \Rightarrow$$

$$\begin{cases} V_k(t,t-m) = V_{k-1}(t,t-m+1)-a_{n+1-k}(t-m)V_n(t,t-m+1) & \text{if } k > 1 \\ V_1(t,t-m) = -a_n(t-m)V_n(t-m+1) \end{cases}$$

This recurrence gives for $V_n(t,t-m)$ two expressions depending on m greater or lower than n.

$$V_n(t,t-m) = -\sum_{i=1}^{n} a_i(t-m)V_n(t-m+i) \quad \text{if } m \geq n$$

If we notice that $V_n(t,t-m)=\Phi(t,t-m)B$ or $h(t,t-m)=C_tV_n(t,t-m)$, we see that the relation (69) is proved, and $q=n-1$. The relation (70) comes from the recurrence relation on $V_n(t,t-m)$ when $m < n$:

$$V_n(t,t-m) = -\sum_{i=1}^{m} a_i(t-m)V_n(t-m+i)+V_m(t,t) \text{ if } m < n$$

and since $\Phi(t,t)=I$, the vector $V_{n-m}(t,t)$ has only one nonzero element located at the (n-m)-th line so that: $C_t V_{n-m}(t,t)=b_m(t,t)$ and (70) follows directly.

Sufficient condition:
Assume that the $a_i(t)$ and $b_i(t)$ satisfy (69) and (70) for the system with impulse response $h(t,s)$. To the input ε_t of this system one can associate the signal w_t:

$$w_t = \varepsilon_t - \sum_{i=1}^{p} a_i(t-i)w_{t-i}$$

One defines the state x_t as:

$$x_t = \begin{bmatrix} w_{t-n+1} \\ . \\ w_{t-1} \\ w_t \end{bmatrix}$$

It is then clear that x_t satisfies the relation (68), and that the system (68) realizes the impulse response $h(t,s)$. ∎

Remarks:

As in the stationary case, one observes immediately that the companion form of the state equation (67) induces the observability of the system, and that (68) implies its controllability.

The relations (61), (63), (69) and (70) are to be seen as relations between functions of the time t, and should be satisfied for every time t, p and q being kept constant.

The comparison of relations (61) and (69) shows that the $a_i(t)$ appearing in the two state equations are not the same. The same is also true for the $b_i(t)$. This is a consequence of the nonstationarity, while for a stationary model, the two canonical forms (observable and controllable) have the same coefficients and the state equations are simply dual of each other. The proof of realizability under the controllable form in the stationary case uses the same quantity w_t issued from the "AR part" or the denominator of the system; but everything is simpler since the polynomials A(z) and B(z), which are the z-transforms of the AR part and MA part respectively, commute.

In the nonstationary case, it would still be possible to use a generalized z-transform (KAMEN, KHARGONEKAR, 1982), but the ring of polynomials defined in that context is not commutative. That is why the previous demonstration could only be done in the time domain, where the transition matrix $\Phi(t,s)$ can be made explicit. If the matrix A was time invariant, the matrix Φ would be a power of A: $\Phi(t,s)=A^{t-s}$, and the Cayley-Hamilton theorem would give directly the companion canonical form. The absence of this theorem in the nonstationary case is a second reason for the difference between the two canonical observable and controllable forms.

6.2.3 Connecting ARMA model and canonical controllable form.

We have seen that the observable form is equivalent to the ARMA model. What about the controllable form? If we examine conditions (63) and (69), and we interpret them, as will be done latter in terms of matrices, the first one will be a condition of rank on the lines of the generalized Hankel or Toëplitz matrix. The other one will be a condition on the columns. There is no reason to believe that the ranks p,q associated with condition (69) and with the controllable form coincide with those of the observable form or the ARMA model. This mimics a result by KAMEN, HAFEZ, 1979, showing that the replacement of an arbitrary (noncanonical) nonstationary state-space equation by the observable canonical form may require an increase in the dimension of the space vector. Therefore, it does not seem possible to give more than a sufficient condition for the controllable form (68) to correspond to an ARMA model.

Proposition.
In order for the model (68) to realize an ARMA model of order p,q with $p \leq n$ and $q < n$, it is sufficient that it is observable.

Demonstration :
In fact what we seek is the autoregressive part of the model, and its coefficients $\alpha_k(t)$ such that the residual has a finite impulse response, $f(t, t-j) = 0$ for $j > q$:

$$\sum_{k=0}^{p} \alpha_k(t-k) y_{t-k} = \sum_{j=0}^{\infty} f(t, t-j) \varepsilon_{t-j}$$

Introducing $h(t, t-i) = C_t A_{t-1} \cdots A_{t-i} B$, the response of the controllable model, in the preceding relation, we obtain :

$$\sum_{k=0}^{p} \alpha_k(t-k) \sum_{i=0}^{\infty} h(t-k, t-k-i) \varepsilon_{t-k-i} = \sum_{j=0}^{\infty} f(t, t-j) \varepsilon_{t-j}$$

Then by equating the coefficients of ε_{t-j} :

$$f(t, t-j) = \sum_{k=0}^{\min(p,j)} \alpha_k(t-k) h(t-k, t-j)$$

In order for the $\alpha_k(t)$ to be the AR part of the model, it is sufficient that they satisfy :

$$\sum_{k=0}^{p} \alpha_k(t-k) C_{t-k} A_{t-k-1} \cdots A_{t-j} B = 0 \quad \text{for } j \geq p$$

Since the system is controllable, the columns of $A_{t-p} \cdots A_{t-j} B$ for $j \geq p$ span the whole state space, and it suffices that the coefficients $\alpha_k(t)$ satisfy :

$$\sum_{k=0}^{p} \alpha_k(t-k) C_{t-k} A_{t-k-1} \cdots A_{t-j} = [0 \cdots 0]$$

If the state-space model is observable, this system of equations admits a solution :

$$[1 \; \alpha_1(t-1) \cdots \alpha_p(t-p)] \begin{bmatrix} C_t A_{t-1} A_{t-2} \cdots A_{t-p} \\ C_{t-1} A_{t-1} A_{t-p+1} \\ \cdots \\ C_{t-p} \end{bmatrix} = [0 \cdots 0]$$

The MA part follows : $(j=0...p-1)$

$$\beta_j(t) = f(t, t-j) = \sum_{k=0}^{j} \alpha_k(t-k) C_{t-k} A_{t-k-1} \cdots A_{t-j} G$$

If the model is not observable, the preceding system giving the $\alpha_k(t)$ may have no solution for $p \le n$. ∎

7. RATIONAL RELIEF.

7.1 Definition of the rational relief.

We can now define a relief starting from any of the three forms encountered : ARMA model, observable state equation, or controllable state equation. In order to make it compatible with the stationary spectrum, it is sufficient to freeze the parameters of the model, as if we were taking a photograph of the model at time t; the relief at this instant will be defined as the spectrum of the stationary signal synthesized by a time-invariant model equal to the model "tangent" at time t to the trajectory of the nonstationary model. Each one of these three possible models leads to a rational relief, in the sense that it is a rational fraction in ω, with the general formulation :

$$(71) \qquad \rho(t, \omega) = \left[\frac{B(t,z) B(t,z^{-1})}{A(t,z) A(t,z^{-1})} \right]_{z=e^{j\omega}}$$

The definition of polynomials $A(t,z)$ and $B(t,z)$ depends on the selected representation (among the three mentioned above), but also on the conventions defining the time indices in the model. We have seen that a nonstationary ARMA model could be written in two different manners (synchronous or shifted). For synchronous models defined by (57), we have :

$$(72) \qquad A(t,z) = \sum_{i=0}^{p} a_i(t-i) z^{-i}$$

while the shifted model (56), leads to

$$(73) \qquad A(t,z) = \sum_{i=0}^{p} a_i(t) z^{-i}$$

The definition $B(t,z)$ is of course parallel to (72) and (73). This distinction between the two definitions is only superficial since if we use in (72) the coefficients of the synchronous form, and in (73) those of the shifted form, the two formulations become identical. Less trivially, we can introduce in (73) the coefficients of the synchronous form, and we obtain a definition which is different from (72) and admits a more formal interpretation. If we define a shift operator z (which is not the z-transform!), such that $z(f(t))=f(t+1)$, the ARMA model in its synchronous form can be written:

(74)
$$\sum_{i=0}^{p} z^{-i}(a_i(t)y_t) = \sum_{j=0}^{q} z^{-j}(b_j(t)\varepsilon_t)$$

This formulation is purely formal and does not coincide with the usual z-transform but is connected to the generalisation given by KAMEN, KHARGONEKAR, 1982. If one freezes the parameters of the model in (74) one obtains again the definition of the relief issued from (73) with (71).

With the model under observable state equation form, the same duality happens. If the equations are frozen at time t, the polynomials become:

(75)
$$\begin{cases} A(t,z) = \sum_{i=0}^{p} a_i(t-1)z^{-i} \\ B(t,z) = \sum_{j=0}^{q} b_j(t)z^{-j} \end{cases}$$

where $p=n$ and $q=n-1$, while in a more formal way, using the shift operator, the state equation becomes:

(76)
$$\begin{cases} x_t = z^{-1}(A_t z_t) + B_t \varepsilon_t \\ y_t = C x_t \end{cases}$$

When this model is frozen, the polynomial $A(t,z)$ has the same definition as in (73), with coefficients $a_i(t)$ and $b_i(t)$ taken from the synchronous version (57) of the ARMA model.

In the controllable form, the same alternative appears, and the polynomial $A(t,z)$ can be written as either:

(77)
$$A(t,z) = \sum_{i=0}^{p} a_i(t-i)z^{-i}$$

or, using the shift operator:

(78)
$$A(t,z) = \sum_{i=0}^{p} a_i(t+1-i)z^{-i}$$

Let us recall that in this last form, the $a_i(t)$ and $b_i(t)$ are not those appearing in the ARMA model, whereas the observable form had the same $a_i(t)$ and $b_i(t)$ as the ARMA model.

The choice between definitions should not consider the controlable form which is not connected to the ARMA model. We should then use (74) or (76) as the models which are to be frozen. These two formulas express the equivalence between the ARMA model and its observable state space realization. The rational relief is therefore defined by (71) or (73).

7.2 Properties of the rational relief.

Let us now describe the properties of this definition, with respect to the four keywords selected in section 2: distribution, nature, locality, invariance. In order to have a proper distribution it is necessary that the marginal distributions be the variance and the spectrum of the signal. This is satisfied, in the frequency domain, by all the variants of the rational relief since the relief is identical to the spectrum when the signal is stationary (A5/B5). This is however, not satisfied in the time domain as the following counter-example demonstrates. Let y_t be an autoregressive nonstationary signal of order 1, such that $y_t + a_1(t-1)y_{t-1} = \varepsilon_t$ with $a_1(t) = -\alpha$ if $t < 0$ or $t > T$, and $a_1(t) = -1/\alpha$ for $0 \leq t \leq T$, $|\alpha| < 1$, α real. Then the variance of y_t is:

$$(79) \qquad e(y_t^2) = \begin{cases} \dfrac{1}{1-\alpha^2} & \text{if } t < 0 \\[3mm] \dfrac{\alpha^2 - 2\alpha^{-2t-2}}{\alpha^2 - 1} & \text{if } 0 \leq t \leq T \\[3mm] \dfrac{1 - \alpha^{2(t-T)}(1+\alpha^2 - 2\alpha^{2T-2})}{1-\alpha^2} & \text{if } t > T \end{cases}$$

The variance is constant up to time 0 after which the instability of the stationary tangent model implies an exponential growth of this variance, then, after time T, the stability of the tangent model induces a decrease of the variance. This variations of the variance appear nowhere on the relief defined by the ARMA model or by the state equations (by the way, for an order 1, there is no distinction between observable form and controllable form).

The nature of this relief is rather satisfactory: the relief is real (A1/B1) and positive (B8/C3). However, it is only weakly connected to the covariance of the signal, since it is computed from the Wold decomposition of the signal, like the relief by Tjøstheim-Mélard, but because of the replacement of the impulse response by an ARMA model or a state-space equation, the relief is no longer the result of a linear transform of the covariance.

The locality of the relief is effective, since the definition was elaborated in order to follow strictly the condition on the slices (A6/B6). The relief of a signal which is composed of a succession of slices of stationary signals is exactly the succession of the corresponding spectra if all the slices are taken from signals having the same innovation ε_t, $t \in (-\infty, +\infty)$, and the relief is composed of the succession of the spectra, except inside a transition interval of duration equal to the order $n = max(p, q+1)$ of each rational slice, in the case where the slices are independent from each other.

The last aspect is the invariance of the relief under filtering or translation. The invariance under time translation (B11) is of course satisfied by each of the rational definitions. The invariance under filtering is not satisfied, like for the relief by Tjøstheim-Mélard, except by approximation: if the parameters $a_i(t)$ and $b_i(t)$ evolve slowly, the relief of the signal filtered by a stationary filter is close to the product of the initial relief by the squared modulus of the transfer function of the filter. The invariance under frequency translation comes from the impulse response $h(t,s)$ of the Wold decomposition of the signal y_t. We can use the argument given by MELARD, 1978-a, according to which the innovation of $y_{0,t}=y_t e^{j\omega_0 t}$ is $\eta_t=\epsilon_t e^{j\omega_0}$, the response $\gamma(t,s)$ associated to $y_{0,t}$ becomes $\gamma(t,s)=h(t,s)e^{j\omega_0 t}$. The invariance under frequency translation of the spectrum defined by (72) from the ARMA model, follows from the expression (80) for the coefficients $a_{0,m}(t)$, $b_{0,m}(t)$ of the model generating $y_{0,t}$ expression which is a consequence of (61) and (63):

$$(80) \qquad \begin{cases} a_{0,m}(t) = a_m(t)e^{j\omega_0 m} \\ b_{0,m}(t) = b_m(t)e^{j\omega_0 m} \end{cases}$$

The same relation holds for the coefficients of the controllable state-space model. It is easy to verify that each of the definitions preserves this invariance under frequency translation in the same manner as the relief by Tjøstheim-Mélard. The property of invariance under time reversal which was not satisfied by the relief by Tjøstheim-Mélard, is not satisfied either by the rational relief. It would be possible to show, but it is of little interest, that this invariance under time reversal is approximately satisfied for every signal with parameters $a_i(t)$ and $b_i(t)$ that evolve slowly.

8. ESTIMATION OF RATIONAL RELIEFS.

Now that we have studied the definition of rational reliefs for nonstationary signals, we can investigate their estimation. Clearly, we should use parametric methods, and estimate time-dependent ARMA models: that was our aim in introducing rational reliefs. This section describes a set of possible approaches, and focuses on one of them, based upon time-varying ARMA models which are parametrized both in frequency (as a finite-order recurrence equation) and in time (through a finite expansion of the time-dependent coefficients over a basis of functions).

8.1 Classes of nonstationary estimators.

When dealing with a nonstationary signal, it is impossible to make an estimation of the value of the instantaneous parameters of the ARMA model without any further assumption. We have to introduce some additional information, either known before the modeling procedure, or based on an initial guess which is verified afterwards. When looking at the litterature on the subject, one can distinguish four main philosophies.

The first approach is based upon the hypothesis that the signal is stationary except at some instants where its generating model jumps from one value (a set of parameters) to another one. The main problem in that case is not the estimation of each stationary model which can be done through classical stationary methods, provided that the length of each interval of stationarity is sufficient. The difficulty is the detection of the jumps: locating them is a non-linear problem which is often advantageously done simultaneously with the model estimation. A review of these methods is given by BENVENISTE, BASSEVILLE, 1986.

A second approach is based upon adaptive estimation methods. The underlying structure of the model is stationary in that case, and the nonstationarity is governed by the algorithms, which are based upon sliding estimation windows or forgetting factor techniques. With these tricks, the estimation of the model at time t simply requires a correction of the model estimated at time $t-1$, according to the new incoming sample(s). These methods are now widely spread within the signal processing community. A reference book is by LJUNG, SODERSTROM, 1983, DURRANI, 1987.

A third approach is based upon time-varying models parametrized in time: the coefficients of these models are time functions which are assumed to be (or are approximated as) linear combinations of a finite number of known functions. The weights of these combinations are the unknown parameters of the models, and are time-invariant, which is the key for their estimation. The main contributions to this approaches were done by LIPORACE, 1975, HALL et al., 1977, in the autoregressive case. A more complete list of references can be found in GRENIER, 1983-b. This methodology will be described more precisely in the following sections.

The fourth approach was investigated in econometrics, but was quite never used in signal processing. It consists in time-varying models with random coefficients. This requires a two-level model: the first level is a description of the signal in terms of its innovation and its generating system, the second level is a model for the evolution of the coefficients in the first level. Usually this second level is a multi-input, multi-output system, its outputs are the first level coefficients, its inputs are noise samples. A short review of that field can be found in the monography by NICHOLLS, QUINN, 1982.

8.2 Time-varying AR models.

The time-varying autoregressive models were first treated by LIPORACE, 1975. Here, the model is slightly modified to give eq. (81), with time-varying coefficients expressed over a basis of functions $f_i(t)$, $i=0..m$.

(81) $$y_t + a_1(t-1)y_{t-1} + ... + a_p(t-p)y_{t-p} = \varepsilon_t \quad \text{with } a_i(t) = \sum_{j=0}^{m} a_{ij}f_j(t)$$

The functions $f_i(t)$ can be powers of time, Legendre polynomials, sines and cosines (truncation of Fourier series)... An interpretation in terms of a vector signal was given by GRENIER, 1983-b. Let Y_t be the vector with $m+1$ components, equal to the product of the nonstationary signal by the vector containing the function basis, and θ be the vector of parameters a_{ij}, then (81) is rewritten as (82):

$$(82) \qquad y_t + [\, a_{10}...a_{1m} \;\; a_{20} \,...a_{pm} \,] \begin{bmatrix} Y_{t-1} \\ \cdot \\ \cdot \\ \cdot \\ Y_{t-p} \end{bmatrix} = \varepsilon_t \;\; \text{or} \;\; y_t + [Y_{t-1}^T \cdots Y_{t-p}^T]\theta = \varepsilon_t$$

$$\text{with} \;\; Y_t = \begin{bmatrix} f_0(t) \\ \cdot \\ \cdot \\ \cdot \\ f_m(t) \end{bmatrix} y_t \;\; \text{and} \;\; \theta = [\, a_{10}...a_{1m} \;\; a_{20} \,...a_{pm} \,]^T$$

This transforms the scalar time-varying model into a vector but time-invariant model. The invariance of the unknown parameters now allows the estimation on a set of samples y_t, $t \in [0,T]$. A least-square estimator is given by eq.(83), where Γ denotes a time interval discussed below.

$$(83) \qquad \sum_{t \in \Gamma} \begin{bmatrix} Y_{t-1} \\ \cdot \\ \cdot \\ \cdot \\ Y_{t-p} \end{bmatrix} [Y_{t-1}^T \cdots Y_{t-p}^T].\theta = -\sum_{t \in \Gamma} \begin{bmatrix} Y_{t-1} \\ \cdot \\ \cdot \\ \cdot \\ Y_{t-p} \end{bmatrix} y_t$$

This system of equations can be solved through fast algorithms, due to the structure of its matrix which is bloc-Toëplitz if Γ is extended to $]-\infty,+\infty[$ by adding zeros to the signal ("correlation method"), or the product of two bloc-Toëplitz matrices if Γ is restricted to the available data : $\Gamma=[p,T]$ ("covariance method"), see HALL et al., 1977.

In this least-square solution, it is assumed that ε_t is a white noise, with unit variance, but it is also possible to extend this method to the case where the residual ε_t is a white noise with time-dependent variance (GRENIER, 1983-b). The most efficient representation for ε_t assumes that it is the product of a white noise (with unit variance) by a gain $b_0(t)$ which is the exponential of a linear combination of the functions in the basis :

$$(84) \qquad b_0(t) = e^{g(t)} \;\; \text{with} \;\; g(t) = \sum_{i=0}^{m} g_i f_i(t)$$

A linearized estimator for the g_i's is built with the logarithm of the absolute value of ε_t :

$$(85) \qquad (\sum_t \begin{bmatrix} f_0(t) \\ \cdot \\ \cdot \\ \cdot \\ f_m(t) \end{bmatrix} [f_0(t) \cdots f_m(t)]) \begin{bmatrix} g_0 \\ \cdot \\ \cdot \\ \cdot \\ g_m \end{bmatrix} = \sqrt{\frac{\pi}{2}} \sum_t \begin{bmatrix} f_0(t) \\ \cdot \\ \cdot \\ \cdot \\ f_m(t) \end{bmatrix} (Log \, |\varepsilon_t| + \frac{C + Log\,2}{2})$$

where C=0.577 215 665 (Euler's constant).

8.3 Time-varying lattices.

The time-varying autoregressive models described above are sufficient for many nonstationary estimation problems. They can also be used for signal synthesis (e.g. speech: CHEVALIER, CHOLLET, GRENIER, 1985). However, especially for this last purpose, they can be replaced by lattice filters which are equivalent, in the stationary case, to AR models. Let $\varepsilon_i^+(p)$ and $\varepsilon_i^-(p)$ be the residuals of a forward and a backward AR model, both of order p. The usual recursion for each cell of the lattice (ITAKURA, SAITO, 1971) is rewritten with time-varying coefficients:

$$(86) \qquad \begin{bmatrix} \varepsilon_i^+(t) \\ \varepsilon_i^-(t) \end{bmatrix} = \begin{bmatrix} 1 & k_i^+(t-1) \\ k_i^-(t) & 1 \end{bmatrix} \begin{bmatrix} \varepsilon_{i-1}^+(t) \\ \varepsilon_{i-1}^-(t-1) \end{bmatrix}$$

Each reflection coefficient $k_i(t)$ is expressed over the function basis with weights k_{ij}^+ and k_{ij}^-. The estimation can be done cell by cell, in a way similar to Burg's algorithm, starting with $\varepsilon_0^+(t)=\varepsilon_0^-(t)=y_t$ (GRENIER, 1983-b). Assuming that the cells 1 to $i-1$ have been estimated, the k_{ij} are estimated by minimizing the energy of the outputs defined by eq.(86), this leads to eq. (87):

$$(87) \qquad \Phi_i^+ . K_i^+ = \Psi_i^+ \quad \text{and} \quad \Phi_i^- . K_i^- = -\Psi_i^-$$

$$\text{with } K_i^+ = \begin{bmatrix} k_{i0}^+ \\ . \\ k_{im}^+ \end{bmatrix} \quad K_i^- = \begin{bmatrix} k_{i0}^- \\ . \\ k_{im}^- \end{bmatrix} \quad H_i^+(t) = \begin{bmatrix} f_0(t)\varepsilon_i^+(t) \\ . \\ f_m(t)\varepsilon_i^+(t) \end{bmatrix} \quad H_i^-(t) = \begin{bmatrix} f_0(t)\varepsilon_i^-(t) \\ . \\ f_m(t)\varepsilon_i^-(t) \end{bmatrix}$$

$$\Phi_i^+ = \sum_{t=i}^{T} H_{i-1}^-(t-1)(H_{i-1}^-(t-1))^T \qquad \Phi_i^- = \sum_{t=i}^{T} H_{i-1}^+(t)(H_{i-1}^+(t))^T$$

$$\text{and} \qquad \Psi_i^+ = \sum_{t=i}^{T} H_{i-1}^-(t-1)\varepsilon_{i-1}^+(t) \qquad \Psi_i^- = \sum_{t=i}^{T} H_{i-1}^+(t)\varepsilon_{i-1}^-(t-1)$$

It is easy to reconstruct the forward and backward AR models from the reflection coefficients, mixing the AR equations with the recursion (86) on the residuals. However, one should notice that the basis upon which the AR coefficients are expressed is an overset of the basis used for the $k_i(t)$.

The AR model, as well as the lattice filter, suffers from a major drawback, which makes synthesis very difficult with these models: the instantaneous values of the model may be unstable over short time intervals. During these intervals, the output of the synthesis filter blows out. Even if the energy falls back to its previous level when the model becomes stable again, this produces bursts in the signal which are audible and sometimes completely degrade the quality of the synthesized signal. A remedy to this defect lies in the replacement of the k_{ij} by another representation for the $k_i(t)$. A suitable one is given by the Log Area Ratios $\gamma_i(t)$, defined by eq. (88), and expressed over the function basis with weights γ_{ij} (OMNES-CHEVALIER, GRENIER, 1988).

(88) $$\gamma_i(t) = Log\left[\frac{1+k_i(t)}{1-k_i(t)}\right] \quad et \quad \gamma_i(t) = \sum_{j=0}^{m}\gamma_{ij}f_j(t)$$

The estimation of the γ_{ij} is done cell by cell, like the k_{ij} of the lattice filter. The estimation of each cell consists into two steps: first estimating reflection coefficients according to eq. (87), and then, approximating the trajectories $\gamma_i(t)$ (obtained from the $k_i(t)$ through eq. (88)), as a linear combination of functions $f_i(t)$. This procedure is a linearized approximation to the least-square estimation which would otherwise lead to a nonlinear problem because of the logarithm in eq. (88). Let $F(t)$ be the vector containing the basis functions: $F(t)=[f_0(t) \cdots f_m(t)]^T$, the second step is solved through eq. (89):

(89) $$\sum_t F(t)F^T(t)\begin{bmatrix}\gamma_{i0}^+ & \gamma_{i0}^- \\ \cdot & \cdot \\ \cdot & \cdot \\ \gamma_{im}^+ & \gamma_{im}^-\end{bmatrix} = \sum_t F(t)\left[Log\left[\frac{1+k_i^+(t)}{1-k_i^+(t)}\right] \quad Log\left[\frac{1+k_i^-(t)}{1-k_i^-(t)}\right]\right]$$

The summation over t is performed for all instants where both forward and backward reflection coefficients have modulus lower than 1.

8.4 Time-varying ARMA models.

The models described above are nonstationary extensions of all-pole models: autoregressive models, lattice filters, and Log Area Ratios, but pole-zero models can also be extended within this framework. Such a model is formulated as in eq. (90) where the $b_i(t)$ are the moving-average coefficients, expressed over the function basis, with weights b_{ij}, and the input ε_t is a unit-variance white noise.

(90) $$y_t+a_1(t-1)y_{t-1}+...+a_p(t-p)y_{t-p} = b_0(t)\varepsilon_t+b_1(t-1)\varepsilon_{t-1}+...+b_q(t-q)\varepsilon_{t-q}$$

A procedure for the estimation of this model was proposed by GRENIER, 1983-b. In a first step, the autoregressive part is estimated through an equation similar to eq. (83), but with instrumental variables (columns), shifted towards the past by q samples, q being the moving-average order. In a second step, the residual which contains only the MA part is estimated with a long AR model. In a third step, this long AR model is inverted to give the MA part: this is done by computing the input of the MA part (filtering the residual of the first step by the inverse AR model from the second step) and estimating the MA part from measurements of its input and its output, which is a linear problem.

A variant of this procedure replaces the second and third step by a nonstationary factorization (GRENIER, 1983-a). The residual from step 1 is multiplied by the basis to give a vector signal. The correlation of this signal, computed as a summation over time is factorized (Schur algorithm). Then the vector model is converted into a scalar one by combining its lines. The weights of this combination are selected to maximize the likelihood of the equality between the estimated correlations, and those deduced from the scalar nonstationary model.

It is also possible to estimate simultaneously the AR and MA part of the ARMA model (GRENIER, 1985-a): first, compute a long time-varying AR model, then determine the a_{ij}

and b_{ij} which minimize the squared error between the (two-indices) impulse responses of the long AR model and the desired ARMA model. Again this is a linearized procedure, which combines the advantages of a good accuracy and a low cost.

8.5 Examples.

Two examples of rational reliefs computed through the previous estimators are presented in this section. Figure 1 shows the relief of the word "zero" in French, uttered by a male speaker. It is clear that the signal is nonstationary, one recognizes the four parts corresponding to the four phonemes in this word. Those familiar with sonagrams will also notice that pitch which is usually seen as periodic vertical bars during the voiced portions of the signal, is not present in this relief, since the model estimates only the vocal tract.

Classical (stationary) linear prediction would require to cut each signal into short windows with a duration of 20-25 ms, located every 10 ms. Here the sampling frequency is 8 kHz, the signal contains about 5200 samples, and lasts 0.65 seconds, this implies 65 windows. For AR(12) models, 780 parameters are needed in the classical linear prediction, while the rational relief presented here uses a time-varying AR(12) model evolving over the 5 first functions in the Fourier basis : 60 parameters are sufficient for the relief.

Figure 2 shows the relief of a nonstationary signal which is an echolocation call (sonar) emitted by a bat, Myotis Mystacinus, during the pursuit of a prey. The complete sequence of calls comprised 55 calls, this one is the 45-th. The signals were recorded by P.Flandrin, ICPI, Lyon, and were sampled at 320 kHz. The call studied here had a duration of 0.45 ms (150 samples), and was modeled through an AR(8) model, with a basis of 4 Legendre polynomials. It shows two components : a frequency-modulated sinusoid, starting at 32 kHz, ending at 25 kHz, with a minimum at 24 kHz, and a second sinusoid, which appears between 0.05 and 0.20 ms, and is a second harmonic of the main component. A more detailed study of these signals can be found in GRENIER, 1986.

9. CONCLUSION.

This text devoted to parametric reliefs, described several definitions. The definition by PRIESTLEY, 1965, which is not satisfactory, is often rejected, so that there remain the definition by TJØSTHEIM, 1976, and the three variants of the rational relief. Among these, the variant based upon controllable state-space equations should also be rejected mainly for practical reasons, since it is difficult or even impossible to determine its value from the sole knowledge of an ARMA model of the signal. The ARMA variant and the observable state variant can be made equivalent by a proper choice of the time origins for each coefficient in the model.

The comparison between the definition by Tjøstheim and the rational definition (observable state equations), leaves them at an equal level : the definition by Tjøstheim preserves the variance of the signal, but the rational definition satisfies the condition of locality by slices. All the other properties are simultaneously present in the two definitions or simultaneously absent.

Parametric estimators for these reliefs were also presented, with an emphasis on time-varying AR and ARMA models, which offer the advantage of including the evolutions of the model with respect to time, through the use of a basis of functions. Examples from selected applications (speech, sonar) confirm that these models are good candidates for the estimation of rational reliefs.

APPENDIX. ELEMENTS OF OPERATOR THEORY.

The object of this appendix is to summarize several elements from the theory of operators which were used in this text, mainly in order to define the vocabulary since the same notions bear different names. These elements come from AKHIEZER, GLAZMAN, 1961, FUHRMANN, 1981, BALAKRISHNAN, 1976, and also the short but very complete book by BEALS, 1971.

Let H be an (infinite dimensional) Hilbert space, with the scalar product $<x,y>$, closed with respect to the norm induced by this scalar product. It is sufficient here to consider only linear operators in H. To the operator A one associates its adjoint A^* such that for all x and y, one has $<x,A^*y>=<Ax,y>$. A normal operator is an operator which commutes with its adjoint. The norm of an operator A is defined by :

$$|A| = \underset{x \neq 0}{Sup} \frac{|Ax|}{|x|}$$

An operator is bounded if its norm is finite. A bounded operator is self-adjoint if $A = A^*$. An operator is unitary if $<Ux,Uy>=<x,y>$ for all x and all y. An operator such that $P^2=P$ and $P^*=P$ is a projection operator.

The most useful concept for the definition of reliefs is the resolution of the identity. A simple example of resolution of the identity is that of a Hermitian matrix (or a symetric matrix in the real case), which represents a self-adjoint operator given a basis in a space H with finite dimension. If T is the matrix, t_i its (real) eigenvalues, and v_i the eigenvector associated to the eigen value t_i, the matrix T can be written :

(A-1)
$$T = \sum_{i=1}^{n} t_i v_i v_i^T$$

With $P_i = \sum_{j=1}^{i} v_j v_j^T$, the matrix T becomes :

(A-2)
$$T = \sum_{i=1}^{n} t_i (P_i - P_{i-1})$$

Here the matrix P_i represents the projection over the subspace spanned by the first i eigenvectors, ordered by decreasing eigenvalues.

When H is of infinite dimension, if T is a compact operator (which transforms any bounded set into a subset of a compact set), also called completely continuous operator, the set of its eigenvalues is countable, and 0 is its single accumulation point. If we define $M_i=\{x, Tx=t_ix\}$ and $M_0=\{x, Tx=0\}$, then P_i (resp. P_0) the projection over M_i (resp. M_0), we can rewrite (A-2) as:

$$\text{(A-3)} \qquad T = \sum_{i=0}^{\infty} t_i P_i \text{ and } I = \sum_{i=0}^{\infty} P_i$$

In this relation, I = identity operator.

A self-adjoint operator T has real eigenvalues. For every real t, define G_t as the subspace spanned by the eigenvectors of T associated with the eigenvalues strictly lower than t, and E_t as the projection over G_t. Then E_{t-0} and E_{t+0} exist and $E_{t-0}=E_t$. If t_k is an eigenvalue, $E_{t_k+0}-E_{t_k-0}=P_k$. This leads to (A-4).

$$\text{(A-4)} \qquad Tx = \int_{-\infty}^{+\infty} t\, dE_t x \text{ and } x = \int_{-\infty}^{+\infty} dE_t x$$

The definition of the resolution of the identity, also called spectral measure is the following one: a resolution of the identity is a one-parameter family E_t, of projection operators, such that:

$$\text{(A-5)} \qquad \begin{cases} t \in [a,b] \ a,b \text{ finite or infinite,} \\ E_{t-0} = E_t \text{ for } a<t<b, \\ E_t E_s = E_{min(t,s)} \\ E_a = 0 \\ E_b = I \end{cases}$$

If U is a unitary operator, the resolution of the identity for U^k is given by:

$$\text{(A-6)} \qquad U^k x = \int_{-\pi}^{+\pi} e^{jkt}\, dE_t x$$

Proof:
To show that, define $c_k=<U^k x, x>$, the sequence of c_k is then positive, and there exists a nondecreasing function $\sigma_x(t)$ such that:

$$\begin{cases} \sigma_x(0) = 0 \\ \sigma_x(t-0) = \sigma_x(t) \\ c_k = \int_{-\pi}^{+\pi} e^{jkt}\, d\sigma_x(t) \end{cases}$$

In order to obtain $<U^k x, y>$, define

$$\sigma(t,x,y) = \frac{1}{4}(\sigma_{x+y}(t)-\sigma_{x-y}(t)+j\sigma_{x+jy}(t)-j\sigma_{x-jy}(t)), \text{ then:}$$

$$\int_{-\pi}^{+\pi} e^{jk}d\sigma(t,x,y) = <U^k x,y>$$

Next, $\sigma(t,x,y)$ being for any t a bilinear functional can be written $\sigma(t,x,y) = <E_t x,y>$ and E_t is a resolution of the identity. ∎

This last result can be extended to a continuous group of unitary operators, i.e. a family
$$<U_t x,y>$$
$$\}$$
right ""
of unitary operators U_t, such that:

The resolution of the identity for the group is:

(A-7)
$$U_t x = \int_{-\infty}^{+\infty} e^{jts}dE_s x$$

The resolution of the identity E_t for a self-adjoint operator T also permits to define functions of T as:

(A-8)
$$\phi(T)x = \int_{-\infty}^{+\infty} \phi(t)dE_t x$$

This relation admits as special cases (A-4) with $\phi(t) = t$ and (A-7) with $\phi_t(s) = e^{jts}$, so that (A-7) can be rewritten as:

(A-9)
$$U_t = e^{jtT}$$

This association of a self-adjoint operator with a continuous group of unitary operators is Stone's theorem, and the operator T associated to the group U_t is called the infinitesimal generator of the group, which can be computed as:

(A-10)
$$T = \lim_{t\to 0}\frac{1}{t}(U_t - U_0)$$

Theorem (Sz-Nagy):
If T is a bounded operator on H, such that $|T^n| \leq k$, there exists a self-adjoint operator Q such that:

$$\frac{1}{k}I \leq Q \leq kI \quad \text{and} \quad QTQ^{-1} \text{ is unitary.}$$

REFERENCES.

N.A.ABDRABBO, M.B.PRIESTLEY, 1967,
"On the prediction of nonstationary processes",
J. of the Royal Statist. Soc., Series B, Vol.29, n° 3, pp 570-585.

N.I.AKHIEZER, I.M.GLAZMAN, 1961,
"Theory of linear operators in Hilbert space",
F.Ungar Publishing Co. (tome 1, 1961, tome 2, 1963).

F.BATTAGLIA, 1979,
"Some extensions of the evolutionary spectral analysis of a stochastic process",
Bull. Unione Matematica Italiana, Vol.16-B, n° 5, pp 1154-1166.

A.V.BALAKRISHNAN, 1976,
"Applied functional analysis",
Springer Verlag.

R.BEALS, 1971,
"Topics in operator theory",
The University of Chicago Press.

A.BENVENISTE, M.BASSEVILLE, 1986,
"Detection of abrupt changes in signals and dynamical systems",
Springer-Verlag.

S.BITTANTI, P.BOLZERN, G.GUARDABASSI, 1985,
"Some critical issues on the state-representation of time-varying ARMA models",
7-th IFAC Symposium on Identification and System Parameter Estimation, York, UK.

A.BLANC-LAPIERRE, 1979,
"Décompositions doublement orthogonales pour des fonctions aléatoires du second ordre
non-stationnaires possédant une puissance moyenne positive",
Colloque GRETSI sur le Traitement du Signal et ses applications, Nice, pp 1/1-1/7.

A.BLANC-LAPIERRE, B.PICINBONO, 1955,
"Remarques sur la notion de spectre instantané de puissance",
Publications Scient. Univ. Alger, Série B, Vol.1, n° 1, pp 17-32.

A.BLANC-LAPIERRE, B.PICINBONO, 1981,
"Fonctions aléatoires",
Masson.

S.BOCHNER, 1956,
"Stationarity, boundedness, almost periodicity of random valued functions",
Proc. 3rd Berkeley Symp. on Math. Statist. and Probability, Univ. of California Press,
pp 7-27.

M.C.CHEVALIER, G.CHOLLET, Y.GRENIER, 1985,
"Speech analysis and restitution using time-dependent autoregressive models",

IEEE-ICASSP 85, Tampa, Florida, pp 501-504.

T.A.C.M.CLAASEN, W.F.G.MECKLENBRAUKER, 1980,
"The Wigner distribution, a tool for time-frequency signal analysis, Part 1 : continuous-time signals, Part 2 : discrete-time signals, Part 3 : relations with other time-frequency signal transformations",
Philips J.Res., Vol.35, n° 3, pp 217-250, pp 276-300, Vol.35, n° 6, pp 372-389.

H.CRAMER, 1961-a,
"On some classes of nonstationary stochastic processes",
Proc. of 4th Berkeley Symp. on Mathematics, Statistics and Probability, Vol.2, Univ. of California Press, pp 57-78.

H.CRAMER, 1961-b,
"On the structure of purely nondeterministic stochastic processes",
Arkiv for Matematik, Vol.4, n° 19, pp 249-266.

H.CRAMER, 1964,
"Stochastic processes as curves in Hilbert space",
Th. of Probability and its appl., Vol.9, n° 2, pp 169-179.

H.CRAMER, 1966,
"A contribution to the multiplicity theory of stochastic processes",
Proc. 5th Berkeley Symp. on Math., Statist. and Prob., Vol.2, pp 215-221.

N.G.DE BRUIJN, 1967,
"Uncertainty principles in Fourier analysis",
Inequalities 2, O.Shisha Ed., Academic Press, pp 57-71

A.DE SCHUTTER-HERTELEER, 1977,
"Une généralisation de concepts spectraux non-stationnaires",
Colloque : Séries chronologiques, aspects fréquentiels et temporels, Cahiers du CERO, Vol.19, n° 3-4, pp 365-377.

T.S.DURRANI, 1987,
"Adaptive Signal Processing",
Signal Processing, J.L.Lacoume, T.S.Durrani, R.Stora Eds., North-Holland, pp 483-513.

B.ESCUDIE, 1979,
"Représentations en temps et fréquence de signaux d'énergie finie : analyse et observation des signaux",
Ann. des Télécom., Vol.34, n° 3-4, pp 101-111.

P.A.FUHRMANN, 1981,
"Linear systems and operators in Hilbert space",
Mc Graw-Hill.

R.K.GETOOR, 1956,
"The shift operator for nonstationary stochastic processes",
Duke Math. J., Vol.23, n°1, pp 175-187.

O.D.GRACE, 1981,
"Instantaneous power spectra",
J. Acoust. Soc. Am., Vol.69, n° 1, pp 191-198.

Y.GRENIER, 1981-a,
"Estimation de spectres rationnels non-stationnaires",
Colloque GRETSI sur le traitement du signal et ses applications, Nice, pp 185-192.

Y.GRENIER, 1981-b,
"Rational nonstationary spectra and their estimation",
First ASSP Workshop on Spectral Estimation, Hamilton, Ontario, pp 6.8.1-6.8.8.

Y.GRENIER, 1983-a,
"Estimation of nonstationary moving-average models",
IEEE ICASSP-83, pp 268-271.

Y.GRENIER, 1983-b,
"Time-dependent ARMA modeling of nonstationary signals",
IEEE Trans. on ASSP, Vol.31, n° 4, pp 899-911.

Y.GRENIER, 1984,
"Modélisation de signaux non-stationnaires",
Thèse de Doctorat d'Etat, Paris.

Y.GRENIER, 1985-a,
"Estimation simultanée AR et MA d'un modèle non-stationnaire",
Colloque GRETSI sur le Traitement du signal et ses Applications, Nice, pp 41-45.

Y.GRENIER, 1986,
"Nonstationary signal modeling with application to Bat echolocation calls",
Acustica, Vol.61, n°3,. pp 155-165.

Y.GRENIER, D.ABOUTAJDINE, 1984,
"Comparaison des représentations temps-fréquence de signaux présentant des discontinuités spectrales".
Annales des Télécommunications, Vol.38, n° 11-12, pp 429-442.

Y.GRENIER, M.C.OMNES-CHEVALIER, 1988,
"Autoregressive models with time-dependent Log Area Ratios",
IEEE Trans. on ASSP, Vol.36, n°10, pp 1602-1612.

M.HALL, A.V.OPPENHEIM, A.WILLSKY, 1977,
"Time-varying parametric modelling of speech",
IEEE Decision and Control Conf., New Orleans, pp 1085-1091.

M.HALL, A.V.OPPENHEIM, A.WILLSKY, 1983,
"Time-varying parametric modelling of speech",
Signal Processing, Vol.5, n° 3, pp 267-285.

M.HALLIN, 1978,
"Mixed autoregressive-moving average multivariate processes with time-dependent coefficients",
J. of Multivariate Analysis, Vol.8, n° 4, pp 567-572.

M.HALLIN, G.MELARD, 1977,
"Indéterminabilité pure et inversibilité des processus autorégressifs moyenne mobile à coefficients dépendant du temps",
Cahiers du CERO, Vol.19, n° 3-4, pp 385-392.

F.ITAKURA, S.SAITO, 1971,
"Digital filtering techniques for speech analysis and synthesis",
7th Int. Cong. Axoustics, Budapest, pp 261-264.

E.I.JURY, 1964,
"Theory and application of the z-transform method",
J.Wiley and sons, NY (pp 59-78).

E.W.KAMEN, K.M.HAFEZ, 1979,
"Algebraic theory of linear time-varying systems",
SIAM J. on Control and Optimization, Vol.17, n° 4, pp 550-510.

E.W.KAMEN, P.P.KHARGONEKAR, 1982,
"A transfer function approach to linear time-varying discrete-time systems",
IEEE CDC-82, pp 152-157.

H.KOREZLIOGLU, 1963,
"Prévision et filtrage linéaire, applications à la représentation canonique des processus aléatoires du second ordre et à la théorie de l'information",
Thèse de Doctorat d'Etat, Paris.

L.A.LIPORACE, 1975,
"Linear estimation of nonstationary signals",
J. Acoust. Soc. Amer., Vol.58, n° 6, pp 1288-1295.

L.LJUNG, T.SODERSTROM, 1983,
"Theory and practice of recursive identification",
MIT Press.

R.M.LOYNES, 1968,
"On the concept of the spectrum for nonstationary processes",
J. of the Royal Statist. Soc., Series B, Vol.30, n° 1, pp 1-30.

M.M.MARTIN, 1968-a,
"Sur diverses définitions du spectre pour des processus non-stationnaires",
Rev. du Cethedec, Vol.5, n° 16, pp 113-128.

M.M.MARTIN, 1968-b,
"Utilisation des méthodes de l'analyse spectrale à la prévision de certains processus non-stationnaires",

Rev. du Cethedec, Vol.5, n° 16, pp 137-148.

W.MARTIN, 1981,
"Line tracking in nonstationary processes",
Signal Processing, Vol.3, n° 2, pp 147-155.

W.MARTIN, 1982,
"Time-frequency analysis of nonstationary processes",
IEEE Int. Symp. on Inform. Theory, Les Arcs.

P.MASANI, 1978,
"Dilatations as propagators of Hilbertian varietes",
SIAM J. Math. Anal., Vol.9, n° , pp414-456.

G.MELARD, 1978-a,
"Propriétés du spectre évolutif d'un processus non-stationnaire",
Ann. Inst. H.Poincaré, Section B, Vol.14, n° 4, pp 411-424.

G.MELARD, 1978-b,
"Theoretical problems with the evolutionary spectrum",
Non publié.

A.G.MIAMEE, H.SALEHI, 1978,
"Harmonizability, V-boundedness and stationary dilatations of stochastic processes",
Indiana Univ. Math. J., Vol.27, n° 1, pp 37-50.

K.S.MILLER, 1969,
"Nonstationary autoregressive processes",
IEEE Trans. on IT, Vol.15, n° 2, pp 315-316.

D.F.NICHOLLS, B.G.QUINN, 1982,
"Random coefficient autoregressive models : an introduction",
Springer-Verlag.

H.NIEMI,1976,
"On the linear prediction problem of certain nonstationary stochastic processes",
Math. Scand., Vol.39, n° 1, pp 146-160.

M.B.PRIESTLEY, 1965,
"Evolutionary spectra and nonstationary processes",
J. of the Royal Statist. Soc., Series B, Vol.27, n° 2, pp 204-237.

T.S.RAO, H.TONG, 1974,
"Linear time-dependent systems",
IEEE Trans. on AC, Vol.19, n° 6, pp 735-737.

B.SZ-NAGY, 1947,
"On uniformly bounded linear operators",
Acta Sc. Math. Szeged, Vol.11, pp 152-157.

D.TJØSTHEIM, 1976-a,
"On random processes that are almost strict sense stationary",
Adv. Appl. Prob., Vol.8, n° 4, pp 820-830.

D.TJØSTHEIM, 1976-b,
"Spectral generating operators for nonstationary processes",
Adv. Appl. Prob., Vol.8, n° 4, pp 831-846.

D.TJØSTHEIM, 1976-c,
"A commutation relation for wide sense stationary processes",
SIAM J. Applied Math., Vol.30, n° 1, pp 115-122.

D.TJØSTHEIM, J.B.THOMAS, 1975,
"Some properties and examples of random processes that are almost wide sense stationary",
IEEE Trans. on IT, Vol.21, n° 3, pp 257-262.

J.VILLE, 1948,
"Théorie et applications de la notion de signal analytique",
Cables et Transmissions, Vol.2, n° 1, pp 61-74.

L.A.ZADEH, 1961,
"Time-varying networks",
Proc. of the IRE, Vol.49, pp 1488-1503.

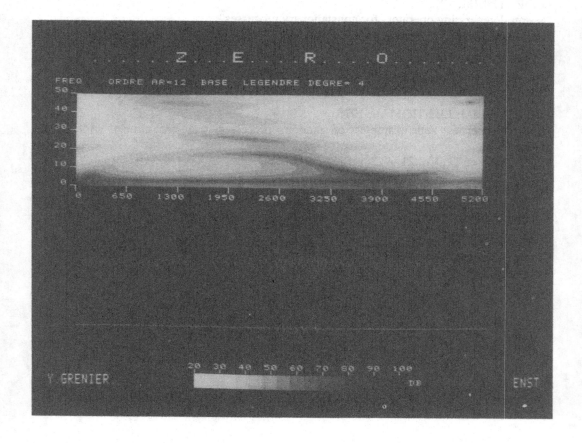

Figure 1. Relief of an utterance of the word "zero" in French, by a male speaker. AR(12) model, Fourier basis, 5 functions. Time is on the horizontal axis, frequency on the vertical axis, grey levels give the energy. Duration: 650 ms, sampling frequency 8 kHz. The upper signal is the original signal, the lower one is the residual error (these signals are undersampled on the display).

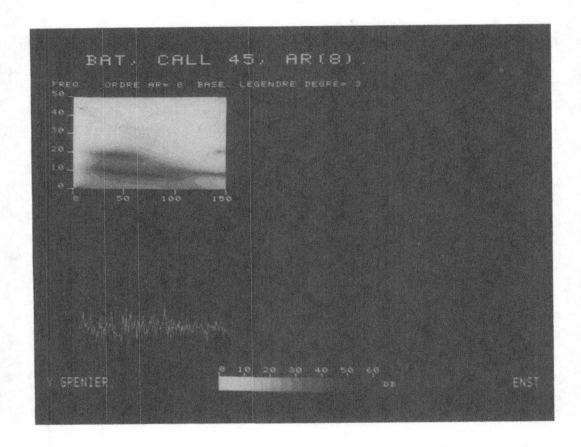

Figure 2. Relief of a bat echolocation call, Myotis Mystacinus, during a pursuit. AR(8) model, 4 Legendre polynomials. Time is on the horizontal axis, frequency on the vertical axis, grey levels give the energy. Duration 0.45 ms, sampling frequency 320 kHz.

Printed in the United States
By Bookmasters